An Introduction to
Polymer
Rheology and
Processing

Nicholas P. Cheremisinoff, Ph.D.
Elastomers Technology Division
Exxon Chemical Company
Linden, New Jersey

CRC Press
Boca Raton   Ann Arbor   London   Tokyo

05182773
CHEMISTRY

Library of Congress Cataloging-in-Publication Data

Cheremisinoff, Nicholas P.
An introduction to polymer rheology and processing / by Nicholas
P. Cheremisinoff.
p. cm.
Includes bibliographical references and index.
ISBN 0-8493-4402-6
1. Polymers--Rheology. I. Title.
TA455.P58C476 1992
620.1'9204292--dc20   92-16516
                                                                              CIP

International Standard Book Number 0-8493-4402-6

Library of Congress Card Number 92-16516

Printed in the United States of America 1 2 3 4 5 6 7 8 9 0

Printed on acid-free paper

PREFACE

This volume is intended as a practical introduction to the subject of rheology with emphasis given to polymer applications. It was written with the practitioner in mind, specifically those individuals requiring a working knowledge of major rheometric techniques available and the practical constitutive equations that can be applied to data analysis.

The subject of rheometry is extremely important to numerous engineering applications. To the author, this subject has played a dominant role over the years in the development of a variety of new elastomeric products. Rheometric techniques on a practical plane help to define the flow properties of materials. For elastomers and plastics, the rheology of a material establishes whether or not the product can be processed, shaped, and formed into a desired article in an efficient and economical manner, while maintaining dimensional stability and high quality. Laboratory-scale studies on the rheological properties of a polymer melt provide information that can be used to predict processing performance of prototype polymers, to provide guidance in developing optimum processing conditions for polymer compounds and blends, to guide in the molecular and compositional design of the polymer, and to help develop compound formulations that enhance end-use performance characteristics. The rheology of a polymer melt is affected not only by the polymer structure (including composition, levels of crystallinity and branching, and molecular weight and molecular weight distribution), but by the compound formulation (type of ingredients, level of ingredient incorporation, and use of plasticizers, oils, etc.) and the specific processing operation and conditions of processing. Hence, the rheology of polymers is of keen interest not only to the product design specialist, but to the compounder, the fabricator, and the engineer. It is a complex subject about which scores of textbooks have been written over the years, many of which address the fascinating and perplexing aspects of the theoretical foundations of the subject.

This volume is not intended as a textbook, nor are theoretical considerations addressed in detail. It is intended as a practical desk reference, providing an overview of operating principles, data interpretation, and qualitative explanation of the importance and relation of rheology to polymer processing operations.

Nicholas P. Cheremisinoff

THE AUTHOR

Nicholas P. Cheremisinoff, Ph.D., is with the Elastomers Technology Division of Exxon Chemical Company in Linden, New Jersey, where he supervises and conducts product development programs on ethylene-propylene-diene monomer rubbers. His industrial and research experiences span a wide range in the rubber and petrochemicals industries and in environmental fields. He received his B.S., M.S., and Ph.D. degrees in Chemical Engineering from Clarkson College of Technology, Potsdam, New York. Dr. Cheremisinoff is the author, co-author, and editor of over 100 engineering textbooks and several hundred articles and research papers and he is the editor of the multivolume *Handbook of Polymer Science and Technology,* Marcel-Dekker Publishers, Basel, Switzerland.

TABLE OF CONTENTS

Chapter 1

INTRODUCTION TO RHEOLOGY

I. GENERAL REMARKS

The subject of rheology is concerned with the study of flow and deformation. From a broad perspective, rheology includes almost every aspect of the study of the deformation of matter under the influence of imposed stress. In other words, it is the study of the internal response of materials to forces. Between the extremes of the conceptual views of the Newtonian fluid and the Hookean solid lie materials of great interest. Commercial interest in synthetic polymeric materials has given the greatest impetus to the science of rheology.

When a small stress is suddenly exerted on a solid, a deformation begins. The material will continue to deform until molecular (internal) stresses are established, which balance the external stresses. The term ''deformation'' refers to the equilibrium deformation that is established when the internal and external stresses are in balance. Most solids exhibit some degree of elastic response, in which there is complete recovery of deformation upon removal of the deforming stresses. The simplest such body is the Hookean elastic solid, for which the deformation is directly proportional to the applied stress. Elastic response may also be exhibited by non-Hookean materials, for which the deformation is not linearly related to the applied stress.

Not all materials reach an equilibrium deformation. In a fluid, if an external stress is exerted, deformation occurs, and continues to occur indefinitely until the stress is removed. A fluid response is one in which no resistance to deformation occurs. Internal frictional forces retard the rate of deformation, however, and an equilibrium can be established in which the rate of deformation is constant and related to the properties of the fluid. The simplest such fluid is the Newtonian, in which the rate of deformation is directly proportional to the applied stress. Many fluids exist which exhibit a nonlinear response to stress and are referred to as non-Newtonian fluids. Most synthetic polymer solutions and melts exhibit some degree of non-Newtonian behavior.

Between the extremes of elastic and fluid response lies a spectrum of combinations of these basic types of material behavior. For example, there is plastic response, wherein a material deforms like an elastic solid as long as the applied stress is below some limit, called the yield stress. If the applied stress exceeds the yield stress the material behaves as a fluid. A common example is paint. Brushing imposes stresses sufficiently large that paint behaves like a fluid. When paint lies on a vertical surface in a thin film, however, the stresses that arise from the weight of the fluid are below the yield stress, and the paint remains on the surface to dry as a uniform film.

1

Another important class of materials is the viscoelastic fluid. Such a material resists deformation, but at the same time resists a time rate of change of deformation. These materials exhibit a combination of both elastic and fluid response.

A materials response to a stress not only depends on the material, but also on the time scale of the experiment. For example, water behaves like a Newtonian fluid in ordinary experiments, but, if subjected to ultrahigh frequency vibrations, it will propagate waves as if it were a solid. The reason for this apparent change in the type of behavior lies in the fact that response is ultimately molecular in nature, and involves the stretching of intermolecular bonds and the motion of molecules past one another. In general, bonds can be stretched very quickly by an imposed stress, since little motion is involved. On the other hand, considerably more time is involved in causing molecules to "flow". Thus, in a stress field with a very short time scale (high frequency vibration is an example) the stress may reverse itself before molecules have time to move appreciably, and only mechanisms giving rise to elasticity may have time to be excited. Because behavior can depend upon the type of stress field imposed, it is important that both the rheological properties of a material as well as the range of conditions over which these properties were measured be recorded. Only in this way can rheological properties be used with any assurance that they have pertinence to the application at hand.

II. CONSTITUTIVE EQUATIONS

The response of any element of a body to the forces acting upon that element must satisfy the principle of conservation of momentum. In a continuum this principle is embodied in the "dynamic equations", written in Cartesian coordinates as

$$\rho \left(\frac{\partial v_i}{\partial t} + v_j \frac{\partial v_i}{\partial x_j} \right) = \rho f_i + \frac{\partial \tau_{ij}}{\partial x_j} \tag{1}$$

for an incompressible fluid of density ρ subject to an external force field f. The stress tensor τ is usually interpreted as being comprised of the mean normal stress $p = -\frac{1}{3}\tau_{ii}$, and the excess over the mean due to dynamic stresses $\bar{\tau}$. For a fluid at rest, $\bar{\tau} = 0$, and the mean normal stress is just the hydrodynamic pressure. Thus, with $\tau = -p\delta + \bar{\tau}$, the dynamic equations become

$$\rho \left(\frac{\partial v_i}{\partial t} + v_j \frac{\partial v_i}{\partial x_j} \right) = \rho f_i - \frac{\partial P}{\partial x_i} + \frac{\partial \tau_{ij}}{\partial x_j} \tag{2}$$

Of course, Equation 2 represents three equations, one for each direction of the coordinate system. Note that repeated subscripts imply summation over that subscript. δ is the Kronecker delta, or unit tensor. The negative sign in front of p corresponds to the convention that a pressure is a negative stress and a tension is a positive stress. By these definitions of p and $\bar{\tau}$, $\tau_{ii} = 0$.

In addition to the dynamic equations and equation that expresses the principle of conservation of mass (known as the continuity equation), we may write for an incompressible material (in Cartesian coordinates)

$$\frac{\partial v_i}{\partial x_i} = 0 \tag{3}$$

Often one wishes to use cylindrical or spherical coordinates in the solution of a problem. In that case, Equations 2 and 3 must be transformed. Standard textbooks on fluid mechanics provide Equations 2 and 3 in Cartesian, cylindrical, and spherical coordinate systems. The number of unknowns in these equations is much greater than the number of equations at hand. Normally f is given in a dynamic problem and p is known. Hence, the unknowns are the three components of v, the pressure p, and the nine components of $\bar{\tau}$. For most real fluids, $\bar{\tau}$ is a symmetric tensor and has only six independent components. Hence, ten unknowns are to be found from only four equations. The additional six equations required are the so-called constitutive equations for the material, which relate the components of τ to the velocity and its derivatives.

For example, a simple constitutive equation is that of the Newtonian fluid:

$$\tau_{ij} = \eta_0 \left(\frac{\partial v_i}{\partial x_j} + \frac{\partial v_j}{\partial x_i} \right) \tag{4}$$

where η_0 is a constant and is known as the coefficient of viscosity.

Once the constitutive equation is established, the dynamic problem is determinate. Unfortunately, it may not be amenable to solution for complex materials (defined by the constitutive equation) or for complex boundary conditions.

Fluids respond to stress by deforming or flowing. Flow is basically a process in which the material deforms at a finite rate. The basic kinematic measure of the response of a fluid is the rate of deformation tensor $\bar{\Delta}$,

$$\Delta_{ij} = \frac{\partial v_i}{\partial x_j} + \frac{\partial v_j}{\partial x_i} = \Delta_{ji} \tag{5}$$

Motion may exist in a fluid even if $\bar{\Delta}$ is identically zero. Each element of fluid may be translating at the same linear velocity, and each element may, in addition, have the same angular velocity about some axis, due to a rigid rotation of the fluid. However, uniform translation and rotation do not contribute to the deformation of the fluid, and so are not associated with that part of the response of the material that is of interest.

In addition to a dynamic response, real fluids exhibit a thermodynamic response. Deformation gives rise to frictional forces within the fluid, and this friction dissipates a part of the kinetic energy of the fluid and causes it to appear as heat. It is possible that sufficient heat is generated to raise the temperature of the fluid appreciably.

This leads one to observe that no flow is isothermal, despite any precautions of thermostating the boundaries of the system. However, for relatively low rates of deformation, the temperature rise is insufficient to change the properties of the fluid. On the other hand, for many important flows, the viscosity of the fluid is so high that even small deformation rates generate significant amounts of heat. In this case one must be able to correct any calculations based on an isothermal analysis. This requires a knowledge of the temperature field in the fluid, as well as a knowledge of the effect of temperature on fluid properties such as viscosity.

The temperature field in an incompressible fluid satisfies an energy equation which, in Cartesian coordinates, has the form

$$\rho \hat{C}_p \left(\frac{\partial T}{\partial t} + v_j \frac{\partial T}{\partial x_j} \right) = \frac{\partial}{\partial x_j} \left(k \frac{\partial T}{\partial x_j} \right) + \tau_{ij} \frac{\partial v_i}{\partial x_j} \tag{6}$$

The heat capacity per unit mass is $\hat{C}_p$ and the thermal conductivity is k. The term gives the volumetric rate of conversion of kinetic energy into heat through friction. It is a dissipation term.

III. SHEAR FLOW

A well-known flow situation is "simple shear flow", in which a nonzero component of velocity occurs only in a single direction. If the subscripts 1, 2, and 3 denote, respectively, the flow direction, the direction of velocity variation, and the neutral direction, then a simple shear flow is defined by

$$v = (v_i \quad , \quad 0 \quad , \quad 0) \tag{7}$$

$$\bar{\Delta} = \dot{\gamma}(x_2) \begin{pmatrix} 0 & 1 & 0 \\ 1 & 0 & 0 \\ 0 & 0 & 0 \end{pmatrix} \tag{8}$$

These flows are sometimes called "viscometric flows", because they are achieved in capillary, Couette, and cone and plate viscometers. Note that in Equation 8, $\dot{\gamma}(X_2)$ is a scalar function of the X_2 coordinate. $\dot{\gamma}$ is a "shear" component of the rate of deformation tensor, and is commonly referred to as "shear rate".

The correspondence of the simple shear flow notation to the more common coordinate notation is given in Figure 1 for some simple flows of interest. The flows shown correspond to the most common flows found in viscometric instruments, but are by no means the only simple shear flows.

IV. CLASSIFICATION OF NON-NEWTONIAN BEHAVIOR THROUGH FLOW CURVES

A plot of τ_{xy} vs. $\dot{\gamma}$ results in a straight line passing through the origin and having slope μ. The resultant plot is known as a flow curve. Fluid materials that display nonlinear behavior through the origin at a given temperature and pressure are non-Newtonian (refer to Figure 2). Non-Newtonian fluids are broadly classified time-independent fluids, time-dependent fluids, and viscoelastic fluids. The first category is comprised of fluids for which the shear rate at any point is solely a function of the instantaneous shear stress. In contrast, time-dependent fluids are those for which shear rate depends on both the magnitude and duration of shear. Some fluids in this second class also show a relationship between shear rate and the time lapse between consecutive applications of shear stress. Less common viscoelastic fluids display the behavior of partial elastic recovery upon the removal of a deforming shear stress.

Time-dependent non-Newtonian fluids can exhibit the property of a yield stress. Yield stress τ_y can be thought of as a minimum stress value that must be exceeded for deformation to occur. In other words, when $\tau_{xy} > \tau_y$ the internal structure of the fluid remains intact, and when $\tau_{xy} > \tau_y$, shearing movement takes place. Flow curves for this type of materials are illustrated in Figure 3B. An idealistic fluid is the Bingham plastic, in which an attempt is made to describe non-Newtonian behavior with a linear constitutive equation. The rheological flow curve in Figure 3B for this fluid can be stated as:

$$\mu_\alpha = \eta + \frac{\tau_y}{\dot{\gamma}} \tag{9}$$

where μ_α is the apparent viscosity of the fluid and is analogous with the Newtonian apparent viscosity

$$\mu_\alpha = \frac{\tau_{xy}}{\dot{\gamma}} \tag{10}$$

The Bingham plastic constitutive equation contains a yield stress, τ_y, and a term η referred to as the plastic viscosity. Equation 9 states that the apparent viscosity decreases with increasing shear rate. In practical terms this means that a value of the apparent viscosity can only be related as a flow property

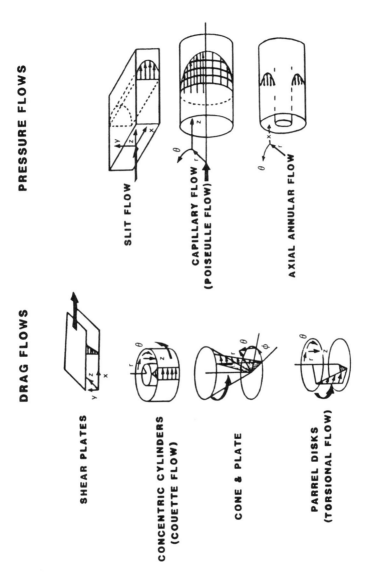

FIGURE 1.   Common shear flow geometries.

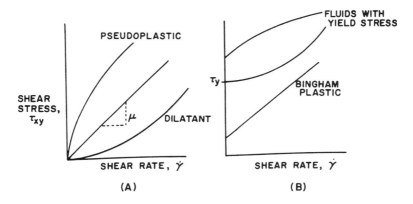

FIGURE 2. Flow curves for non-Newtonian, time-independent materials.

with a corresponding shear rate. Although the Bingham plastic fluid itself is idealistic, it can be applied to modeling a portion of a real non-Newtonian's flow curve. Also, as shown in subsequent chapters, many materials show only a small departure from exact Bingham plasticity and therefore can be approximately described by Equation 9.

A large variety of industrial fluid-like materials may be described as being pseudoplastic. The flow curve is illustrated in Figure 2. In examining this curve, note that it is characterized by linearity at very low and very high shear rates. The slope of the linear region of the curve at the high shear rate range is referred to as the viscosity at infinite shear (μ_0), whereas the slope in the linear portion near the origin is the viscosity at zero shear rate (μ_0). A logarithmic plot of τ_{xy} vs. γ is found to be linear over a relatively wide shear rate range and hence may be described by a power law expression (known as the Ostwald-de Waele model):

$$\tau_{xy} = K\dot{\gamma}^n \tag{11}$$

The slope η and intercept K are given the names "flow behavior index" (or "pseudoplasticity index") and "consistency index", respectively. The power law exponent ranges from unity to zero with increasing plasticity (i.e., at n = 1, the equation reduces to the constitutive equation of a Newtonian fluid). The value of the consistency index is obtained from the intercept on the τ_{xy} axis and hence represents the viscosity at unit shear rate. As shown later for a variety of polymers, K is very sensitive to temperature, whereas n is much less sensitive. By analogy to Newton's law, the apparent viscosity of a power law fluid is

$$\mu_\alpha = K\dot{\gamma}^{n-1} \tag{12}$$

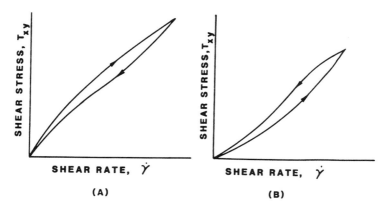

FIGURE 3.   Hysteresis loops for time-dependent fluids. (A) Thixotropic hysteresis; (B) rheopectric hysteresis.

Since n < 1, the apparent viscosity of a pseudoplastic fluid decreases with increasing shear rate, and hence these materials are often referred to as shear thinning.

The next class of time-independent fluids are dilatant materials. Volumetric dilatancy refers to the phenomenon whereby an increase in the total fluid volume under application of shear occurs. Rheological dilatancy refers to an increase in apparent viscosity with increasing shear rate. The flow curve is illustrated in Figure 2. As in the case of a pseudoplastic fluid, a dilatant material is usually characterized by zero yield stress. Hence, the power law model may also be used to describe this fluid behavior, but with n-values greater than unity.

Time-dependent non-Newtonian fluids are classified as thixotropic and rheopectic. Thixotropic fluids display a reversible decrease in shear stress with time at a constant shear rate and temperature. The shear stress of such a material usually approaches some limiting value. Both thixotropic and rheopectic fluids show a characteristic hysteresis as illustrated by the flow curves in Figure 3. The flow curves are constructed from data generated by a single experiment in which the shear rate is steadily increased from zero to some maximum value and then immediately decreased toward zero. The arrows on the curves denote the chronological progress of the experiment. An interesting complexity of these materials is that the hysteresis is time history-dependent. In other words, changing the rate at which γ is increased or decreased in the experiment alters the hysteresis loop. For this reason, generalized approaches to defining an index of thixotropy have not met with much success.

Rheopectic fluids are sometimes referred to as antithixotropic fluids because they exhibit a reversible increase in shear stress over time at a constant rate of shear under isothermal conditions. The location of the hysteresis loop shown in Figure 3B is also dependent on the time history of the material, including the rate at which γ is changed.

TABLE 1
Examples of Non-Newtonian Materials

Time-independent fluids	
Fluids with a yield stress	Various plastic melts, oil well drilling mids, oils, sand suspensions in water, coal suspensions, cement and rock slurries, peat slurries, margarine, shortenings, greases, aqueous thorium oxide slurries, grain water suspensions, toothpaste, soap, detergent slurries, paper pulp
Pseudoplastic fluids (without yield stress)	Rubber solution, adhesives, polymer solutions and melts, greases, starch suspensions, cellulose acetate, mayonnaise, soap, detergent slurries, paper pulp, paints, certain pharmaceutical suspensions, various biological fluids
Dilatant fluids (without yield stress)	Aqueous suspensions of titanium dioxide, corn flour/sugar solutions, gum arabic in water, quicksand, wet beach sand, deflocculated pigment dispersions with suspended solids such as mica and powdered quartz, iron powder dispersed in low-viscosity liquids.
Time-dependent fluids	
Thixotropic fluids	Melts of high molecular weight polymers, some oil well drilling mids, various greases, margarine and shortening, printing inks, various food substances, paints
Rheopectic fluids	Betonite clay suspensions, vanadium pentoxide suspensions, gypsum suspensions, various soils, dilute suspensions of ammonium oleate
Viscoelastic fluids	Bitumens, flour dough, napalm, various jellies, polymers and polymer melts (e.g., nylon), various polymer solutions

The final category of non-Newtonians is viscoelastic fluids, which are materials exhibiting both viscous and elastic properties. Viscoelasticity is described in some detail later. It should be kept in mind that for purely Hookean elastic solids the stress corresponding to a given strain is time-independent, but with viscoelastic materials the stress dissipates over time. Viscoelastic fluids undergo deformation when subjected to stress; however, part of their deformation is gradually recovered when the stress is removed. Therefore, viscoelastic substances may be fancifully referred to as fluids that have memory. Viscoelasticity is frequently observed in the processing of various polymers and plastics. For example, in the production of synthetic fibers such as nylon for ultrafine cable, material is extruded through a die consisting of fine perforations. In these examples the cross-section of the fiber or cable may be considerably larger than that of the perforation through which it was extruded. This behavior is a result of the partial elastic recovery of the material. Table 1 provides examples of the various types of non-Newtonian fluids.

This book attempts to relate rheological characteristics observed in laboratory studies to commercial processing problems. The materials discussed here are largely polymer melts and solutions, and blends of various industrial non-Newtonian materials widely used in industry. These materials display the properties of pseudoplasticity, exhibit normal forces in excess of hydrostatic pressure in simple shear flows, and often display viscoelastic behavior (i.e., stress relaxation, creep recovery, and stress overshoot). Each of these rheological features is discussed separately, but the reader should bear in mind that many materials show all three characteristics. Engineering practices have not yet advanced to the point at which a single generalized constitutive equation may be developed to describe the entire range of rheological features. Clearly, the more features that are incorporated into a constitutive equation, the more complex it becomes, and hence the more difficult it is to apply its use in a balance equation needed to solve a problem in fluid mechanics.

In this monograph the author relies heavily on the use of rheological flow curves to understand and project the processing characteristics of non-Newtonian materials. Some general features to be kept in mind about these flow curves for later discussions are as follows: First, flow curves, especially for many types of polymers, display a Newtonian plateau or region, then a transition to non-Newtonian, and finally a power law region. The upper limit of shear rate of the Newtonian plateau depends on the molecular weight and temperature of the polymer.

This upper limit for many polymers is typically 10^{-2} s^{-1} (nylon and PET are exceptions). This limit decreases with molecular weight and molecular weight distribution, as well as with decreasing temperature. The onset of the transition region between the Newtonian plateau and power law can be related through the dimensionless Deborah number:

$$N_{DEB} = \lambda/temp = \lambda\dot{\gamma} \tag{13}$$

where λ is the material's relaxation time.

The transition region is usually quite distinct for monodispersed polymeric melts, and broad for polydisperse melts. The magnitude of viscosity for each of the regions depends on both material and state variables, as illustrated later. A final characteristic of pseudoplastic flow curves to keep in mind is that the power law index n is in fact not an absolute constant for a specific material. For many materials, however, it often reaches some constant value for a reasonably wide range of conditions.

V. CONSTITUTIVE EQUATIONS FOR PSEUDOPLASTICITY

The constitutive equation for a power law fluid is stated in Equation 17, where apparent viscosity depends on two parameters, K and n. The consistency index K has units of $(N - s^n/m^2)$. K is temperature sensitive and is found to follow the Arrhenius relationship:

TABLE 2
Values of Power Law Model Coefficients for Various Materials

Material	Temp (K)	Shear rate (s^{-1})	K (10^3) $(n - s^2/m^2)$	n
High-density polyethylene (HDPE)	453	100–1000	6.19	0.56
HDPE	473	100–1000	4.68	0.59
Alathon (TM) 7040	473	100–6000	4.31	0.47
(E. I. Dupont de Nemours & Co.)				
Low-density polyethylene (LDPE)	433	100–4000	9.36	0.41
LDPE	453	100–6500	5.21	0.46
Alathon (TM) 1540	473	100–6000	4.31	0.47
(E. I. Dupont de Nemours & Co.)				
Polypropylene (PP)	483	100–3000	32.1	0.25
E 612 (TM) (Exxon Chemical Co.)	513	50–3000	22.4	0.28
Nylon	498	100–2500	2.62	0.63
Capron 8200 (TM)	503	100–2000	1.95	0.66
(Allied Chemical Corporation	508	100–2300	1.81	0.66
Polymethyl methacrylate (PMMA)	493	100–6000	88.3	0.19
PMMA	513	100–6000	42.7	0.25
Lucite 147 (TM)	533	100–7000	26.2	0.27
(E. I. Dupont de Nemours & Co.)				

$$K = K_0 \exp\left[\frac{\Delta E}{R_G} \left(\frac{1}{T} - \frac{1}{T_0}\right)\right] \qquad (14)$$

where ΔE = activation energy of flow per mole; R_G = gas constant per mole; T, T_0 = absolute and reference temperatures; and K_0 = constant, characteristic of the material. For relatively small temperature ranges, Equation 14 can be approximated by

$$K = K_0 \exp[-\alpha(T - T_0)] \qquad (15)$$

It should be noted that because of thermodynamic considerations K also depends exponentially on pressure.

The apparent viscosity of a power law fluid is a function of all the velocity gradients in non-simple shearing flows. However, from the standpoint of practicality most laboratory viscometric techniques measure viscosity in only one velocity gradient. This is a general fluid dynamic problem treated in a later chapter. An important point to realize is that the power law equation is an empirical model and therefore should not be used for extrapolation of viscosity data. It does not level off to a Newtonian plateau, but instead continues to increase. Table 2 provides values of the power law coefficients K and n for different materials.

The so-called Carreau model provides a constitutive equation which in fact does level off to a limiting viscosity η_0, which is either measured or

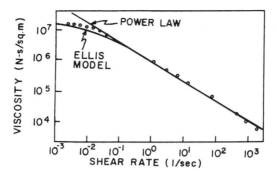

FIGURE 4.   Comparison of power law and Ellis models.

estimated. It is, however, a more complex equation having additional model parameters:

$$\frac{\eta(\dot\gamma) - \eta_\infty}{\eta_0 - \eta_\infty} = [1 - (\lambda\dot\gamma)^2]^{(n-1)/2} \tag{16}$$

η_∞ is the solvent viscosity for solutions. For polymer melts it is common practice for $\eta_\infty = 0$. Note that we have changed our notation for apparent viscosity to the more conventional symbol η.

Another model incorporating the Newtonian plateau viscosity (η_0) is the Ellis model:

$$\frac{\eta}{\eta(\tau)} = \left(\frac{\tau}{\tau_{1/2}}\right)^{\alpha-1} \tag{17}$$

where

$$\eta\big|_{\tau=\tau_{1/2}} = \frac{\eta_0}{2} \tag{18}$$

The exponent (α - 1) is the slope obtained from a plot of (η_0/η - 1) vs. log ($\tau/\tau_{1/2}$).

The Ellis model is sometimes more conveniently expressed as

$$\tau_{xy} = \frac{1}{A + B\tau_{xy}^{\alpha-1}} \dot\gamma \tag{19}$$

where coefficients A and B have units of $L^2\tau^{-1}F^{-1}$ and $L^{2\alpha}\tau^{-1}F^{-\alpha}$, respectively. Term α is dimensionless. Figure 4 provides a comparison of the power law and Ellis models for ABS at two temperatures. Note that the power law model provides an adequate fit of the data for a large position of the shear rate range, but departs from measured values at low shear rates.

REFERENCES

1. Viscoelastic behavior of crystalline polymers; **M. Takayanagi,** *Proc. 4th Int. Congr. on Rheology,* Part 1 (1965), pp. 161–187.
2. The viscoelastic properties of oriented nylon 66 fibers. III. Stress relaxation and dynamic mechanical properties; **T. Murayama, J. H. Dumbleton, and M. L. Williams,** *J. Macromol. Sci.,* B1 (1967), pp. 1–14.
3. On the dynamic mechanical behavior of drawn poly(ethylene terephthalate) fibers; **J. H. Dumbleton, T. Murayama, and J. P. Bell,** *Kolloid-Z. Z. Polym.,* Band 228, Heft 1-2 (1968), pp. 54–58.
4. Effects of orientation on dynamic mechanical properties of ABS; **J. H. Daane and S. Matsuoka,** *Polym. Eng. Sci.,* 8 (4) (Oct. 1968), pp. 246–251.
5. The effect of structural changes on dye diffusion in poly(ethylene terephthalate); **J. H. Dumbleton, J. P. Bell, and T. Murayama,** *J. Appl. Polym. Sci.,* 12 (1968), pp. 2491–2508.
6. Viscoelasticity of oriented poly(ethylene terephthalate); **T. Murayama, J. H. Dumbleton, and M. L. Williams,** *J. Polym. Sci.,* Part A-2, Vol. 6 (1968), pp. 787–793.
7. Dynamic mechanical properties of textile fibres; **R. Meredith,** *Proc. 5th Int. Congr. on Rheology,* Vol. 1 (1969), pp. 43–60.
8. Mechanical anisotropy of crystalline polymers in relation to molecular orientation; **H. Kawai,** *Proc. 5th Int. Congr. on Rheology,* Vol. 1 (1969), pp. 97–131.
9. Relaxations in polyurethanes in the glass transition region; **T. Kajiyama and W. J. MacKnight,** *ACS Polym. Prepr.,* 10 (1) (April 1969), pp. 65–71.
10. Rheological properties of styrene-ethylene oxide block copolymers; **P. F. Erhardt, J. J. O'Malley, and R. C. Crystal,** *ACS Polym. Prepr.,* 10 (2) (Sept. 1969), pp. 812–819.
11. Relations between dynamic mechanical properties and melting behavior of nylon 66 and poly(ethylene terephthalate); **J. P. Bell and T. Murayama,** *J. Polym. Sci.,* Part A-2, Vol. 7 (1969), pp. 1059–1073.
12. Relation between the network structure and dynamic mechanical properties of a typical amine-cured epoxy polymer; **T. Murayama and J. P. Bell,** *J. Polym. Sci.,* Part A-2, Vol. 8 (1970), pp. 437–445.
13. Strain waves in twisted-viscoelastic filament assemblies; **C. F. Zorowski, T. Murayama, and A. T. Alptekin,** *Proc. 5th Int. Congr. on Rheology,* Vol. 3 (1970), pp. 295–310.
14. Antiplasticization in crosslinked polymers; **J. Kumanotani, J. Tsuchiya, K. Shibayama, and M. Kodama,** *Proc. 5th Int. Congr. on Rheology,* Vol. 3 (1970), pp. 431–440.
15. Effect of diluents and crosslinking on secondary mechanical dispersion of polyalkyl methacrylate; **K. Shibayama, T. Tanaka, and M. Kodama,** *Proc. 5th Int. Congr. on Rheology,* Vol. 3 (1970), pp. 451–459.
16. A chemorheological study of the crosslinking reaction in oil-modified phenolic resins; **Y. Takahashi, S. Naganuma, K. Onozawa, and H. Kishi,** *Proc. 5th Int. Congr. on Rheology,* Vol. 3 (1970), pp. 485–494.
17. Mechanical relaxation of single crystal mats of polyethylene in a crystalline dispersion region; **S. Manabe, A. Sakoda, A. Katada, and M. Takayanagi,** *J. Macromol. Sci.,* (March 1970), pp. 161–184.
18. Dynamic mechanical properties of polyurethane block polymers; **D. S. Huh and S. L. Cooper,** *Polym. Eng. Sci.,* 11 (5) (Sept. 1971), pp. 369–376.

19. Bonding characterization in reinforced composites; **C. F. Zorowski and T. Murayama,** *Proc. Int. Conf., on Mechanical Behavior of Materials,* Kyoto, Vol. 5 (Aug. 1971), pp. 28–41.

20. Low temperature dynamic mechanical properties of polyurethane-polyether block copolymers; **J. L. Illinger, N. S. Schneider, and F. E. Karasz,** *Polym. Eng. Sci.,* 12 (1) (Jan. 1972), pp. 25–29.

21. Structure-property relationships of chlorinated polyethylenes; **N. K. Kalfoglou and H. L. Williams,** *Polym. Eng. Sci.,* 12 (3) (May 1972), pp. 224–235.

22. Viscoelastic properties and stability of BBL ladder polymers; **R. P. Chartoff and J. W. Powell,** *30th Annu. Tech. Conf. Soc. of Plastics Engineers,* Chicago (May 1972), pp. 528-532.

23. Steady flow and dynamic viscosity of branched butadiene-styrene block copolymers; **G. Kraus, F. E. Naylor, and K. W. Rollmann,** *Rubber Chem. Techol.,* 45 (4) (June 1972), pp. 1005–1014.

24. Latex interpenetrating polymer networks; **L. H. Sperling, T. W. Chiu, C. P. Hartman, and D. A. Thomas,** *ACS Polym. Prepr.,* 13 (2) (Aug. 1972), pp. 705–709.

25. Morpholoyg, sorption characteristics, and mechanical properties of a cellulose acetate polyacrylonitrile graft; **M. A. Siahkolah and W. K. Walsh,** *ACS Polym. Prepr.,* 13 (2) (Aug. 1972), pp. 716–722.

26. The melting and dynamic mechanical properties of reconstituted collagen; **A. Nguyen, B. T. Vu, and G. L. Wilkes,** *ACS Polym. Prepr.,* 13 (2) (Aug. 1972), pp. 1003–1006.

27. Some morphological factors on thermomechanical analysis of crystalline polymers; **M. Takayanagi,** *ACS Polym. Prepr.,* 13 (2) (Aug. 1972), pp. 1158–1163.

28. Iodine swelling of polyacrylonitrile: evidence for a two-phase structure; **R. D. Andrews, K. Miyachi, and R. S. Doshi,** *ACS Polym. Prepr.,* 13 (2) (Aug. 1972), pp. 1168–1174.

29. Dynamic mechanical and thermal properties of the stratum corneum layer of epidermis; **G. L. Wilkes,** *ACS Polym. Prepr.,* 13 (2) (Aug. 1972), pp. 1189–1192.

30. Structure and properties of crystalline polychloroprene. I. Linear viscoelastic behavior of crystalline/amorphous blends; **R. D. Andrews and N. Kawasaki,** *ACS Polym. Prepr.,* 13 (2) (Aug. 1972), pp. 1200–1205.

31. Dynamic spring analysis; **S. Naganuma, T. Sakurai, Y. Takahashi, and S. Takahashi,** *Chem. High Polym.,* 29 (322) (1972), pp. 105–109.

32. Application of DSA (dynamic spring analysis); **S. Naganuma, T., Sakurai, Y. Takahashi, and S. Takahashi,** *Chem. High Polym.,* 29 (327) (1972) pp. 519-523.

33. Copolymerization of caprolactam with polyoxybutylene diamine; **S. W. Shalaby, H. K. Reimschuessel, and E. M. Pearce,** *Polym. Eng. Sci.,* 13 (2) (March 1973), pp. 88–95.

34. Viscoelastic properties of heterophase blends of homopolymers and block copolymers; **G. Choi, A. Kaya, and M. Shen,** *Polym. Eng. Sci.,* 13 (3) (May 1973), pp. 231–235.

35. The mechanical properties of ultra-oriented polyethylene; **N. E. Weeks and R. S. Porter,** *31st Annu. Tech. Conf. Soc. of Plastics Engineers,* Montreal, (May 1973), pp. 438–440.

36. A study of mechanical and optical properties of alternating and random copolymers of acrylonitrile and butadiene; **D. P. Mukherjee and C. Goldstien,** *ACS Polym. Prepr.,* 14 (1) (May 1973), pp. 36–41.

37. Viscoelastic and friction characteristics of polydiene and butadiene-styrene copolymer tread compounds; **J. L. White and Y. Ming Lin,** *ACS Polym. Prepr.,* 14 (1) (May 1973), pp. 114–119.

38. Physical characterization of an α-methylstyrene-butadiene-styrene block copolymer; **P. S. Pillai and G. S. Fielding-Russell,** *ACS Polym. Prepr.,* 14 (1) (May 1973), pp. 346–351.

39. Effects of water on the mechanical, dielectric, and swelling behavior of a glass-sphere-filled epoxy resin; **J. A. Manson and E. H. Chiu,** *ACS Polym. Prepr.,* 14 (1) (May 1973), pp. 469–474.

40. A bubble inflation technique for the measurement of viscoelastic properties in equal biaxial extensional flow. II; **D. D. Joye, G. W., Poehlin, and C. D. Denson,** *Trans. Soc. Rheol.,* 17 (2) (1973), pp. 287–302.

41. The influence of crystallinity on the β-transition in polyvinylchloride; **E. R. Harrell and R. P. Chartoff,** *ACS Polym. Prepr.,* 14 (2) (Aug. 1973), pp. 755–759.

42. Processing-morphology-property studies of polyvinyl chloride; **C. S. Yusek, T. M. Stephenson, D. M. Gezovick, P. K. C. Tsou, P. H. Geil, and E. A. Collins,** *ACS Polym. Prepr.,* 14 (2) (Aug. 1973), pp. 841–846.

43. Viscoelastic behavior of styrene/*n*-butyl methacrylate/potassium methacrylate polymers; **P. F. Erhardt, J. M. O'Reilly, W. C. Richards, and M. W. Williams,** *ACS Polym. Prepr.,* 14 (2) (Aug. 1973), pp. 902–907.

44. Crystalline, thermal and mechanical properties of block and random copolymers of pivalolactone and (D,L) α-methyl-α-*n*-propyl-β-propiolactone; **A. E. Allegrezza, Jr., R. W. Lenz, J. Cornibert, and R. H. Marchessault,** *ACS Polym. Prepr.,* 14 (2) (Aug. 1973), pp. 1232–1237.

45. Dynamic mechanical properties of some model aromatic polyesters and polycarbonates; **D. J. Massa and J. R. Flock,** *ACS Polym. Prepr.,* 14 (2) (Aug. 1973), pp. 1249–1253.

46. The effect of diluents on molecular mobility in glassy bisphenol A polycarbonate; **S. E. B. Petrie, J. R. Flick, L. J. Garfield, A. S. Marshall, and V. D. Papanu,** *ACS Polym. Prepr.,* 14 (2) (Aug. 1973), pp. 1254–1259.

47. Effect of zinc chloride on nylon 6. II. Dynamic mechanical properties; **M. J. Mehta and R. D. Andrews,** *ACS Polym. Prepr.,* 14 (2) (Aug. 1973), pp. 1260–1264.

48. Effects of gamma radiation on the dynamic mechanical and swelling behavior of styrene butadiene block and random copolymers; **S. L. Samuels and G. L. Wilkes,** *Polym. Eng. Sci.,* 13 (4) (July 1973), pp. 280–286.

49. Chlorinated polyethylene modification of blends derived from waste plastics. II. Mechanism of modification; **C. E. Locke and D. R. Paul,** *Polym. Eng. Sci.,* 13 (4) (July 1973), pp. 308–318.

50. Viscoelastic behavior of butadiene-acrylonitrile copolymers at small and large deformations and their ultimate properties; **N. Nakajima, H. H. Bowerman, and E. A. Collins,** *Rubber Chem. Technol.,* 46 (2) (June 1973), pp. 417–424.

51. Hysteretic losses in rolling tires; **P. R. Willett,** *Rubber Chem. Technol.,* 46 (2) (June 1973), pp. 425–441.

52. Measuring the dynamic moduli of glassy polymers: analysis of the rheovibron; **D. J. Massa,** *J. Appl. Phys.,* 44 (6) (June 1973), pp. 2595–2600.

53. Multiple glass transitions in butadiene-acrylonitrile copolymers. II. Formation of incompatible phases during copolymerization; **A. H. Jørgensen, L. A. Chandler, and E. A. Collins,** *Rubber Chem. Technol.,* 46 (4) (Sept. 1973), pp. 1087–1102.

54. A study of mechanical and optical properties of an alternating NBR and an emulsion NBR; **D. P. Mukherjee and C. Goldstein,** *Rubber Chem. Technol.,* 46 (5) (Dec. 1973), pp. 1264–1273.

55. Synthesis and elastomeric properties of copolymers of butadiene and 1,3-pentadiene; **A. Carbonaro, V. Zamboni, G. Novajra, and G. Dall'Asta,** *Rubber Chem. Technol.,* 46 (4) (Sept. 1973), pp. 1274–1284.

56. Dynamic loss energy measurement of tire cord adhesion to rubber; **T. Murayama and E. L. Lawton,** *J. Appl. Polym. Sci.,* 17 (1973), pp. 669–677.

57. Dynamic response of viscoelastic solids in liquid media; **D. L. MacLean and T. Murayama,** *Trans. Soc. Rheol.,* 18 (2) (1974), pp. 237–245.

58. Properties of compatible blends of poly(vinylidene fluoride) and poly(methyl methacrylate); **D. R. Paul and J. O. Altamirano,** *ACS Polym. Prepr.,* 15 (1) (April 1974), pp. 409–414.

59. Relaxation behavior of blends of homopolymers with block copolymers; **U. Mehra, G. Choi, K. Biliyar, and M. Shen,** *ACS Polym. Prepr.,* 15 (1) (April 1974), pp. 426–431.

60. The effect of molecular conformation on the material properties of poly-(γ-methyl-D-glutamate); **Y. Mohadger, G. L. Wilkes, and C. Anderson,** *ACS Polym. Prepr.,* 15 (1) (April 1974), pp. 593–598.

61. Dynamic mechanical testing of polymers; **G. P. Koo,** *Plast. Eng.* (May 1974), pp. 33–38.

62. Dynamic non-destructive bond strength characterization in non-woven fabrics; **C. Zorowski and T. Murayama,** *presented at 1974 Fiber Soc. Meet.,* Williamsburg, VA.

63. Dynamic mechanical properties of a polypentenamer and its hydrogenated derivatives; **K. Sanui, W. J. MacKnight, and R. W. Lenz,** *Macromolecues,* 7 (1) (Jan.-Feb. 1974), pp. 101–105.

64. Elastic and high tenacity poly(butene-1) fibers; **C. L. Rohn,** *32nd Annu. Tech. Conf. Soc. of Plastics Engineers,* San Francisco (May 1974), pp. 390–391.

65. The influence of crystallinity on the beta-transition in poly(vinyl chloride); **E. R. Harrell, Jr. and R. P. Chartoff,** *Polym. Eng. Sci.,* 14 (5) (May 1974), pp. 362–365.

66. Processing-morphology-property studies of poly(vinyl chloride); **C. Singleton, J. Isner, D. M. Gezovich, P. K. C. Tsou, P. H. Geil, and E. A. Collins,** *Polym. Eng. Sci.,* 14 (5) (May 1974), pp. 371–381.

67. New apparatus for the immersion of test sample in a liquid medium on rheovibron viscoelastometer; **T. Murayama and A. A. Armstrong, Jr.,** *J. Polym. Sci.,* 12 (1974), pp. 1211–1213.

68. Heat generation in tires due to the viscoelastic properties of elastomeric components; **P. R. Willett,** *Rubber Chem. Technol.,* 47 (2) (June 1974), pp. 363–375.

69. Orientation, dynamic mechanical, and thermal behavior of a semi-crystalline block co-polymer; **J. C. West, A. Lilaonitkul, S. L. Cooper, U. Mehra, and M. Shen,** *ACS Polym. Prepr.,* 15 (2) (Sept. 1974), pp. 191–196.

70. Pure head-to-head polymers; **P. Kincaid, T. Tanaka, and O. Vogl,** *ACS Polym. Prepr.,* 15 (2) (Sept. 1974), pp. 222–226.

71. Influence of the structure and morphology on the dynamic mechanical properties of co-polymers 1-hexene-propene; **A. Piloz, A. Douillard, J. Y. DeCroix, and J. F. May,** *XXIII Int. Symp. on Macromolecules,* Madrid, (Sept. 1974), pp. 802–805.

72. Thermal and mechanical transitions in polyethylene terephthalate; **A. M. Munné and G. M. Guzmán,** *XXIII Int. Symp. on Macromolecules,* Madrid (Sept. 1974), pp. 825–829.

73. Thermo-optical, differential calorimetric, and dynamic viscoelastic transitions in poly(2,6-dimethyl-1, 4-phenylene oxide) (PPO[lb] resin) blends with poly-ρ-chlorostyrene and with styrene-ρ-chlorostyrene statistical copolymers; **A. R. Shultz and B. M. Beach,** *Macromolecules,* 7 (6) (Nov.-Dec. 1974), pp. 902–909.

74. Adaptation of the rheovibron to the measurement of dynamic properties of vulcanizates at elongations up to sample rupture; **A. Voet and J. C. Morawski,** *Rubber Chem. Technol.,* 47 (4) (Sept. 1974), pp. 758–764.

75. Dynamic mechanical and electrical properties of vulcanizates at elongations up to sample rupture; **A. Voet and J. C. Morawski,** *Rubber Chem. Technol.,* 47 (4) (Sept. 1974), pp. 765–777.

76. Dynamic viscoelastic properties of raw butadiene-acrylonitrile elastomers; **N. Nakajima, E. A. Collins, and P. R. Kumler,** *Rubber Chem. Technol.,* 47 (4) (Sept. 1974), pp. 778–787.

77. Noise damping with methacrylate/acrylate latex interpenetrating polymer networks; **L. H. Sperling, J. A. Grates, J. E. Lorenz, and D. A. Thomas,** *ACS Polym. Prepr.,* 16 (1) (April 1975), pp. 274–279.

78. A master curve for amorphous elastomers derived from small and large deformations using various instruments; **N. Nakajima and E. A. Collins,** *Rubber Chem. Technol.,* 48 (1) (Mar.-Apr. 1975), pp. 69–78.

79. An apparatus for the measurement of dynamic mechanical properties of polymers in a gas medium with the vibron viscoelastometer; **T. Murayama and B. Silverman,** *J. Appl. Polym. Sci.,* 19 (1975), pp. 1695–1700.

80. Analysis of the rheovibron and its application to the physical characterization of polymers; **D. J. Massa, J. R. Flick, and S. E. B. Petrie,** *ACS Coat. Plast. Prepr.,* 35 (1) (April 1975), pp. 371–376.

81. Morphological characterization of polyester-based elastoplastics; **R. W. Seymour, J. R. Overton, and L. S. Corley,** *Macromolecules,* 8 (3) (May-June 1975), pp. 331–335.

82. Effect of molecular weight on fatigue crack propagation in PMMA; **S. L. Kim, M. Skibo, J. A. Manson, and R. W. Hertzberg,** *ACS Polym. Prepr.,* 16 (2) (Aug. 1975), pp. 559–563.

83. Dilatometric and NMR data related to the γ relaxation of ultra-high molecular weight linear polyethylene; **C. L. Beatty and M. F. Froix,** *ACS Polym. Prepr.,* 16 (2) (Aug. 1975), pp. 628–632.

84. Poly(alkyl α-chloroacrylates). VI. Transitions and relaxations; **G. R. Dever, F. E. Karasz, W. J. MacKnight, and R. W. Lenz,** *Macromolecules,* 8 (4) (July-Aug. 1975), pp. 439–443.

85. Relationships between the dynamic mechanical properties of anisotropic polypropylene films and molecular orientation; **J. C. Seferis, R. L. McCullough, and R. J. Samuels,** *ACS Coat. Plast. Prepr.,* 35 (2) (Aug. 1975), pp. 210–214.

86. Dehydrochlorination of poly(vinyl chloride); **E. P. Chang and R. Salovey,** *Polym. Eng. Sci.,* 15 (8) (Aug. 1975), pp. 612–614.

87. The mechanical behavior of springy polypropylene; **S. L. Cannon, W. O. Statton, and W. J. S. Hearle,** *Polym. Eng. Sci.,* 15 (9) (Sept. 1975), pp. 633–645.

88. Relaxations in cyano-substituted polypentenamers and their hydrogenated derivatives; **R. Neumann, K. Sanui, and W. J. MacKnight,** *Macromolecules,* 8 (5) (Sept.-Oct. 1975), pp. 665–671.

89. Dynamic mechanical and thermal properties of fire-retardant polypropylene; **E. P. Chang, R. Kirsten, and R. Salovey,** *Polym. Eng. Sci.,* 15 (10) (Oct. 1975), pp. 697–702.

90. A new method for the measurement of dynamic shear mechanical properties of materials with a rheovibron viscoelastometer; **T. Murayama,** *J. Appl. Polym. Sci.,* 19 (1985), pp. 3221–3224.

91. Role of polymer science in developing materials for phonograph discs; **S. K. Khanna,** presented at the 52nd Conv. Audio Engineering Soc., Preprint #1060 (B-5) (Nov. 1975).

92. Viscoelastic properties of homogeneous triblock copolymers of styrene-α-methylstyrene and their polyblends with homopolymers; **D. R. Hansen and M. Shen,** *Macromolecules,* 8 (6) (Nov.-Dec. 1975), pp. 903–909.

93. Characterization and piezoelectric activity of stretched and poled poly-(vinylidene flouride). I. Effect of draw ratio and poling conditions; **R. J. Shuford, A. F. Wilde, J. J. Ricca, and G. R. Thomas,** *Polym. Eng. Sci.,* 16 (1) (Jan. 1976), pp. 25–35.

94. Rheovibron determination of the dynamic moduli of polymer melts; **B. H. Shah and R. Darby,** *Polym. Eng. Sci.,* 16 (1) (Jan. 1976), pp. 25–35.

95. Silicone-polyethylene blends; **J. R. Falender, S. E. Lindsey, and J. C. Saam,** *Polym. Eng. Sci.,* 16 (1) (Jan. 1976), pp. 54–58.

96. Linear and nonlinear viscoelastic measurements, ultimate properties, and processability of raw elastomers; **N. Nakajima and E. A. Collins,** *Trans. Soc. Rheol.,* 20 (1) (Jan.-Mar. 1976), pp. 1–21.

97. Tensile strength of polyurethane and other elastomeric block copolymers; **T. L. Smith,** *Rubber Chem. Technol.,* 49 (1) (Mar.-Apr. 1976), pp. 64–84.

98. Mechanical properties of poly(phenylene oxide)/poly(styrene) blends; **A. F. Yee,** *ACS Polym. Prepr.,* 17 (1) (April 1976), pp. 145–150.

99. A molecular mechanism for aging in high density polyethylene; **P. D. Frayer, P. P. Tong, and W. W. Dreher,** *34th Annu. Tech. Conf. Soc. of Plastic Engineers,* Atlantic City, N.J (April 1976), pp. 61–63.

100. Constitutive relationships for the mechanical properties of anisotropic, partialy crystalline polymers; **J. C. Seferis, R. L. McCullough, and R. J. Samuels,** *Polym. Eng. Sci.,* 16 (5) (May 1976), pp. 334–343.

101. Annealing effect on the microstructure of poly(vinyl chloride); **S. Ohta, T. Kajiyama, and M. Takayanagi,** *Polym. Eng. Sci.,* 16 (7) (July 1976), pp. 465–472.

102. Heat generation in pneumatic tires; **P. Kainradl and G. Kaufmann,** *Rubber Chem. Technol.,* 49 (3) (July-Aug. 1976), pp. 823–861.

103. Automation of the rheovibron; **A. S. Kenyon, W. A. Grote, D. A. Wallace, and M. J. Rayford,** *ACS Polym. Prepr.,* 17 (2) (August. 1976), pp. 7–12.

104. Relationships between structure, mobility and physical properties in aromatic polycarbonates and polyesters; **D. J. Massa and P. P. Rusanowsky,** *ACS Polym. Prepr.,* 17 (2) (Aug. 1976), pp. 184–189.

105. Morphology and mechanical behavior of SBS/PS blends; **G. Akovali, M. Niinomi, J. Diamant, and M. Shen,** *ACS Polym. Prepr.,* 17 (2) (Aug. 1976), pp. 560–563.

106. Correlation of H-bonds via DTA, DSC and IR to physical properties of polyether urethaneurea segmented copolymers; **A. R. Cain, W. R. Conrad, S. E. Schonfeld, and B. H. Werner,** *ACS Polym. Prepr.,* 17 (2) (Aug. 1976), pp. 580–584.

107. Properties of organotin polyesters crosslinked by cycloaliphatic epoxides; **R. V. Subramanian and M. Anand,** *ACS Coat. Plast. Prepr.,* 36 (2) (Aug.-Sept. 1976), pp. 233–238.

108. Determination of curing mechanism by dynamic mechanical testing; **C. N. Merriam, T. Ginsberg, and L. M. Robeson,** *ACS Coat. Plast. Prepr.,* 36 (2) (Aug.-Sept. 1976), pp. 358–362.

109. Dynamic mechanical and thermal properties of fire retardant high impact polystyrene; **E. P. Chang, E. L. Slagowski, and R. Kirsten,** *ACS Coat. Plast. Prepr.,* 36 (2) (Aug.-Sept. 1976), pp. 490–496.

110. Prediction of polyethylene melt rheological properties from molecular weight distribution data obtained by gel permeation chromatography; **B. H. Shah and R. Darby,** *Polym. Eng. Sci.,* 16 (8) (Aug. 1976), pp. 579–584.

111. Transition behavior of poly(vinylidene flouride)/poly(ethyl methacrylate) blends; **R. L. Imken, D. R. Paul, and J. W. Barlow,** *Polym. Eng. Sci.,* 16 (9) (Sept. 1976), pp. 593–601.

112. Mechanical properties of biaxially oriented poly(vinyl chloride); **T. E. Brady,** *Polym. Eng. Sci.,* 16 (9) (Sept. 1976), pp. 638–644.

113. A new method for the measurement of dynamic compression mechanical properties of materials with a rheovibron viscoelastometer; **T. Murayama,** *Proc. VIIth Int. Congr. on Rheology,* Sweden (Aug. 1976), pp. 402–403.

114. Antiplasticization in the epoxy resin system; **J. Kumanotani, T. Koshio, and N. Hata,** *Proc. VIIth Int. Congr. on Rheology,* Sweden (Aug. 1976), pp. 540–541.

115. Dielectric and mechanical relaxations in polyphenylquinoxalines; **W. J. Wrasidlo,** *ACS Polym. Prepr.,* 11 (2) (Sept. 1970), pp. 1159–1167.

116. Thermal analysis of aromatic polymers: motions in polyquinoxalines; **W. J. Wrasidlo,** *ACS Polym. Prepr.,* 12 (1) (March 1971), pp. 755–765.

117. Dynamic mechanical properties of anisotropic polypropylene films; **J. C. Seferis, R. L. McCullough, and R. J. Samuels,** *Appl. Polym. Symp.* 27 (1975), pp. 205–228.

118. The role of intercrystalline links in the environmental stress cracking of high density polyethylene; **P. D. Frayer, P. Po-Luk Tong, and W. W. Dreher,** *Polym. Eng. Sci.,* 17 (1) (Jan. 1977), pp. 27–31.

119. Fatigue crack propagation in poly(methyl methacrylate): effect of molecular weight and internal plasticization; **S. L. Kim, M. Skibo, J. A. Manson, and R. W. Hertzberg,** *Polym. Eng. Sci.,* 17 (3) (March 1977), pp. 194–203.

120. Mechanical properties of mixtures of two compatible polymers; **A. F. Yee,** *Polym. Eng. Sci.,* 17 (3) (March 1977), pp. 213–219.

121. Castor oil based interpenetrating polymer networks IV. Mechanical behavior; **G. M. Yenwo, L. H. Sperling, J. Pulido, and J. A. Manson,** *Polym. Eng. Sci.,* 17 (4) (April 1977), pp. 251–256.

122. New block copolymers of α-amino acids; **F. Uralil, T. Hayashi, J. M. Anderson, and A. Hiltner,** *Polym. Eng. Sci.,* 17 (8) (Aug. 1977), pp. 515–522.
123. The influence of diblock copolymers on the structure and properties of polybutadiene-polyisoprene blends; **A. R. Ramos and R. E. Cohen,** *Polym. Eng. Sci.,* 17 (8), (Aug. 1977), pp. 639–646.
124. Bisphenol-A-polycarbonate-bisphenol-A-polysulfone block copolymers; **J. E. McGrath, T. G. Ward, E. Shchori, and A. J. Wnuk,** *Polym. Eng. Sci.,* 17 (8) (Aug. 1977), pp. 647–651.
125. Chlorosulfonated block copolymers; **C. E. Rogers and M. J. Covitch,** *Polym. Eng. Sci.,* 17 (8) (Aug. 1977), pp. 652–656.
126. Effects of post-curing of unsaturated polyester resins in various medias; **S. A. Chen and P. K. Tasai,** *Polym. Eng. Sci.,* 17 (11) (Nov. 1977), pp. 775–781.
127. Dynamic mechanical properties of fibers in liquid media in relation to dyeing; **E. L. Lawton and T. Murayama,** *J. Appl. Polym. Sci.,* 20 (1976), pp. 3033–3055.
128. Mechanical motions in amorphous and semi-crystalline polymers; **R. F. Boyer,** *Polymer,* 17 (Nov. 1976), pp. 996–1008.
129. Properties of polyether-polyester thermoplastic elastomers; **A. Lilaonitkul and S. L. Cooper,** *Rubber Chem. Technol.,* 50 (1) (Mar.-Apr. 1977), pp. 1–23.
130. Compatibilization of rubber blends through phase interaction; **R. F. Bauer and E. A. Dudley,** *Rubber Chem. Technol.,* 50 (1) (Mar.-Apr. 1977), pp. 35-42.
131. Processability of rubber and rheological behavior; **J. L. White,** *Rubber Chem. Technol.,* 50 (1) (Mar.–Apr. 1977), pp. 163–185.
132. Effect of adsorption on reinforcement of filled polyurethane; **J. Seto,** *Rubber Chem. Technol.,* 50 (2) (May–June 1977), pp. 333–341.
133. The role of morphology and structure of carbon blacks in the electrical conductance of vulcanizates; **W. F. Verhelst, K. G. Wolthuis, A. Voet, P. Ehrburger, and J. B. Donnet,** *Rubber Chem. Technol.,* 50 (4) (Sept.-Oct. 1977), pp. 735–746.
134. Fundamental observation on deformation and failure characteristics of SBR 1500; **N. Nakajima and E. A. Collins,** *Rubber Chem. Technol.,* 50 (4) (Sept.-Oct. 1977), pp. 791–797.
135. Polyester-polycarbonate blends; **D. R. Paul, J. W. Barlow, C. A. Cruz, R. N. Mohn, T. R. Nassar, and D. C. Wahrmund,** *ACS Coat. Plast. Prepr.,* 37 (1) (March 1977), pp. 130–135.
136. The effect of cross-linking on the dynamic mechanical properties of an incompatible segmented polyurethane; **J. L. Work, J. E. Herweh, and C. A. Glotfelter,** *ACS Coat. Plast. Prepr.,* 37 (1) (March 1977), pp. 490–495.
137. Effect of annealing on the morphology and properties of thermoplastic polyurethanes; **C. H. M. Jacques,** *ACS Coat. Plast. Prepr.,* 37 (1) (March 1977), pp. 507-512.
138. The dependence of the dynamic mechanical properties of poly[(1,4-cyclohexane) bismethylene isophthalate]-b-polycaprolactone copolymers on segment size and crystallinity; **J. E. Herweh, J. L. Work, and W. Y. Whitmore,** *ACS Coat. Plast. Prepr.,* 37 (1) (March 1977), pp. 724–729.
139. Application of dynamic mechanical testing to coatings research & development; **M. B. Roller and J. K. Gillham,** *ACS Coat. Plast. Prepr.,* 37 (2) (Aug.-Sept. 1977), pp. 135–141.
140. The influence of morphology on semi-crystalline deformation behavior; **C. L. Beatty, J.-C. Pollet and W. J. Stauffer,** *ACS Coat. Plast. Prepr.,* 37 (2) (Aug.-Sept. 1977), pp. 562–570.
141. New block copolymers from α-amino acids; **T. Hayashi, A. Hiltner, and J. M. Anderson,** *ACS Polym. Prepr.,* 18 (1), (March 1977), pp. 229–234.
142. Diblock copolymers as stabilizing agents in elastomer blends; **A. R. Ramos and R. E. Cohen,** *ACS Polym. Prepr.,* 18 (1) (March 1977), pp. 335–339.

143. Polycarbonate-polysulfone block copolymers; **J. E. McGrath, T. C. Ward, E. Shchori, and W. Wnuk,** *ACS Polym. Prepr.,* 18 (1) (March 1977), pp. 346–351.

144. Chlorosulfonated block copolymers; **M. Covitch and C. E. Rogers,** *ACS Polym. Prepr.,* 18 (1) (March 1977), pp. 352–355.

145. Morphology and physical properties of polybutylene terephthalate; **E. P. Chang and E. L. Slagowski,** *ACS Polym. Prepr.,* 18 (1) (March 1977), pp. 635–640.

146. Melt rheology of an ethylene-methacrylic acid copolymer and its sodium salt; **T. R. Earnest, Jr. and W. J. MacKnight,** *ACS Polym. Prepr.,* 18 (2) (Aug. 1977), pp. 391–396.

147. A dynamic relaxation study of a hexafluoroisobutylene/vinylidene flouride copolymer (HF1B/VF$_2$); **M. J. Froix, J. M. Pochan, A. O. Goedde, and T. Davidson,** *ACS Polym. Prepr.,* 18 (2) (Aug. 1977), pp. 450–455.

148. Morphological and physical property effects for solvent cast films of poly-2,5(6) benzimidazole; **A. Wereta, Jr., M. T. Gehatia, and D. R. Wiff,** *Polym. Eng. Sci.,* 18 (3) (Feb. 1978), pp. 204–209.

149. Mechanical properties of films based on novel epoxy/polyamide emulsion; **S. C. Misra, J. A. Manson, and J. W. Vanderhoff,** *ACS Org. Coat. Plast. Chem. Prepr.,* 38 (March 1978), pp. 213–218.

150. Time/temperature cure behavior of epoxy based structural adhesives; **H. M. Li, M. J. Doyle, and A. F. Lewis,** *ACS Org. Coat. Plast. Chem. Prepr.,* 38 (March 1978), pp. 290–295.

151. Dynamic spring analysis; **G. A. Senich, R. Neumann, and W. J. MacKnight,** *ACS Org. Coat. Plast. Chem. Prepr.,* 38 (March 1978), pp. 360–365.

152. The impact fracture toughness of PTFE; **M. Kisbenyi, M. W. Birch, J. M. Hodgkinson, and J. C. Williams,** *ACS Org. Coat. Plast. Chem. Prepr.,* 38 (March 1978), pp. 394–399.

153. A dynamic mechanical study of the curing reaction of two expoxy resins; **G. A. Senich, W. J. MacKnight, and N. S. Schneider,** *ACS Org. Coat. Plast. Chem. Prepr.,* 38 (March 1978), pp. 510–515.

154. A dynamic mechanical study of phase segregation in toluene diisocyanate block polyurethanes; **G. A. Senich and W. J. MacKnight,** *ACS Polym. Prepr.,* 19 (1) (March 1978), pp. 11–16.

155. Morphology and dynamic viscoelastic behavior of blends of styrene-butadiene block copolymers; **G. Kraus, L. M. Fodor, and K. W. Rollmann,** *ACS Polym. Prepr.,* 19 (1) (March 1978), pp. 68–74.

156. Viscoelastic properties of blends and diblock copolymers of polybutadiene and polyisoprene; **A. R. Ramos and R. E. Cohen,** *ACS Polym. Prepr.,* 19 (1) (March 1978), pp. 87–91.

157. Transport properties of chlorosulfonated block copolymers; **H. Y. Chung, M. J. Covitch, and C. E. Rogers,** *ACS Polym. Prepr.,* 19 (1) (March 1978), pp. 98–101.

158. Physical property characteristics of poly(arylene ether)-poly(aryl carbonate) block copolymers; **T. C. Ward, A. J. Wnuk, A. R. Henn, S. Tang, and J. E. McGrath,** *ACS Polym. Prepr.,* 19 (1) (March 1978), pp. 115–120.

159. Physical properties of blends of poly(vinyl chloride) and a terpolymer of ethylene; **H. E. Blair, E. W. Anderson, G. E. Johnson, and T. K. Kwei,** *ACS Polym. Prepr.,* 19(1) (March 1978), pp. 143–148.

160. Physical properties of blends of polystyrene with poly(methyl methacrylate) and styrene/(methyl methacrylate) copolymers; **D. J. Massa,** *ACS Polym. Prepr.,* 19 (1) (March 1978), pp. 157–162.

161. Segmental orientation, physical properties and morphology of poly-ε-caprolactone blends; **D. S. Hubbell and S. L. Cooper,** *ACS Polym. Prepr.,* 19 (1) (March 1978), pp. 163–168.

162. Thermoset polyurethanes containing hydroquinone Di-(β-Hydroxyethyl) ether; **S. A. Iobst and H. W. Cox,** *ACS Polym. Prepr.,* 19 (1) (March 1978), pp. 674–678.

163. Automation of the rheovibron; **A. S. Kenyon, W. A. Grote, D. A. Wallace, and McC. Rayford,** *J. Macromol. Sci. Phys.,* B13 (4) (1977), pp. 553–570.

164. The calorimetric, dynamic mechanical and dielectric properties of polyoxymethylene copolymers; **L. P. DeMejo, W. J. MacKnight, and O. Vogl,** *Soc. of Plastics Engineers 36th Annu. Tech. Conf.,* Vol. XXIV (April 1978), pp. 5–8.

165. Environmental aging of polymeric materials: determination of long-term mechanical properties; **C. F. Pratt,** *Soc. of Plastics Engineers 36th Annu. Conf.,* Vol. XXIV (April 1978), pp. 14–17.

166. Morphology and properties of thermoplastic elastomers; **A. Lilaonitkul, J. W. C. Van Bogart, and S. L. Cooper,** *Soc. of Plastics Engineers 36th Annu. Conf.,* Vol. XXIV (April 1978), pp. 249–253.

167. Factors influencing the impact strength of high impact polystyrene; **E. P. Chang and A. Takahashi,** *Polym. Eng. Sci.,* 18 (5) (April 1978), pp. 350–354.

168. Importance of the massa correction for loss tangent measurements on the rheovibron; **A. R. Ramos, F. S. Bates, and R. E. Cohen,** *J. Polym. Sci.,* 16 (1978), pp. 753–758.

169. Dynamic viscoelastic behavior of ABA block polymers and the nature of the domain boundary; **G. Kraus and K. W. Rollmann,** *J. Polym. Sci.,* 14 (1976), pp. 1133–1148.

170. Viscoelastic behavior of a butadiene-styrene block copolymer containing free polybutadiene molecules; **G. Kraus and K. W. Rollmann,** *J. Polym. Sci.,* Vol. 15 (1977), pp. 385–388.

171. The entanglement plateau in the dynamic modulus of rubbery styrene-diene block copolymers. Significance to pressure-sensitive adhesive formulations; **G. Kraus and K. W. Rollmann,** *J. Appl. Polym. Sci.,* 21 (1977), pp. 3311–3318. (Also refer to *Rubber Chem. Technol.,* 52 (2) (1979), pp. 278–285.)

172. Morphology and viscoelastic behavior of styrene-diene block copolymers in pressure sensitive adhesives; **G. Kraus, F. B. Jones, O. L. Marrs, and K. W. Rollman,** *J. Adhes.,* 8 (1977), pp. 235–258.

173. Homogenous and heterogenous blends of polybutadiene, polyisoprene, and corresponding diblock copolymers; **R. E. Cohen and A. R. Ramos,** *Macromolecules,* 12 (Jan.-Feb. 1979), pp. 131–134.

174. Observation of super-glass-transition event in the damping spectra of amorphous block copolymers; **J. M. G. Cowie and I. J. McEwen,** *Macromolecules,* 12 (Jan.-Feb. 1979), pp. 56–61.

175. Dynamic mehcanical behavior of polystyrene-(ethene-co-butene)-polystyrene, triblock copolymer films cast from various solvents; **J. M. G. Cowie, D. Lath, and I. J. McEwen,** *Macromolecules,* 12 (Jan.-Feb. 1979), pp. 52–56.

176. The measurement of dynamic flexile properties of fibers with a rheovibron viscoelastometer; **T. Murayama,** *J. Appl. Polym. Sci.,* 23 (1979), pp. 1647–1651.

177. Mechanical relaxation processes in polyacrylonitrile polymers and copolymers; **A. S. Kenyon and McC. J. Rayford,** *J. Appl. Polym. Sci.,* 23 (1979), pp. 717–725.

178. Nonlinear viscoelastic behavior of butadiene-acrylonitrile copolymers filled with carbon black; **N. Nakajima, H. H. Bowerman, and E. A. Collins,** *Rubber Chem. Technol.,* 51 (2) (May-June 1978), pp. 322–334.

179. Effect of carbon black on dynamic properties of rubber vulcanizates; **A. I. Medalia,** *Rubber Chem. Technol.,* 51 (3) (July-Aug. 1978), pp. 437–523.

180. Young's Modulus and dynamic mechanical properties of san resins modified with ethylene-propylene rubber; **A. Brancaccio. L. Gargani, and G. P. Giuliani,** *Rubber Chem. Technol.,* 51 (4) (Sept.-Oct. 1978), pp. 655–667.

181. Crystallizable trans-butadiene-piperylene elastomers; **M. Bruzzone, A. Carbonaro, and L. Gargani,** *Rubber Chem. Technol.,* 51 (5) (Nov.-Dec. 1978), pp. 907–924.

182. Effect of crosslink density distribution on some properties of epoxies; **S. C. Misra, J. A. Manson, and L. H. Sperling,** *ACS Org. Coat. Plast. Chem. Prepr.,* 39 (Sept. 1978), pp. 146–151.

183. A study of crosslinking in polyimides by viscoelastic and diffusion techniques; **R. P. Chartoff**, *ACS Org. Coat. Plast. Chem. Prepr.,* 39 (Sept. 1978), pp. 129–135.

184. Dielectric and mechanical studies of sulfonate ionomers; **R. M. Neumann, W. J. MacKnight, and R. D. Lundberg**, *ACS. Polym. Prepr.,* 19 (2) (Sept. 1978), pp. 298–303.

185. Properties of sulfonated polypentenamer ionomers; **D. Rahrig and W. J. MacKnight,** *ACS Polym. Prepr.,* 19 (2) (Sept. 1978), pp. 314–319.

186. Melt rheology of ethylene ionomers; **T. R. Earnest, Jr. and W. J. MacKnight,** *ACS Polym. Prepr.,* 19 (2) (Sept. 1978), pp. 383–387.

187. Effect of molecular weight distribution on fatigue crack propagation in poly(methyl methacrylate); **S. L. Kim, J. Janiszewski, M. D. Skibo, J. A. Manson, and R. W. Hertzberg,** *Polym. Eng. Sci.,* 19 (2) (Mid-Feb. 1979), pp. 145–150.

188. A dynamic mechanical study of the curing reaction of two epoxy resins; **G. A. Senich, W. J. MacKnight, and N. S. Schneider,** *Polym. Eng. Sci.,* 19 (4) (March 1979), pp. 313–318.

189. Preparation of high modulus fibers by zone-annealing under high stress: application to pet and nylon 6; **T. Kunugi, A. Suzuki, I. Akiyama, and M. Hashimoto,** *ACS Polym. Prep.,* 20 (1) (April 1979), pp. 778–779.

190. Preparation of ultra-high modulus polypropylene by zone-drawing apparatus equipped with cooling element and its superstructure; **K. Yamada, M. Kamezawa, and M. Takayanagi,** *ACS Polym. Prepr.,* 20 (1) (April 1979), pp. 780–783.

191. The effect of additives on impact poly(vinyl chloride); **E. P. Chang, R. O. Kirsten, and R. Salovey,** *ACS Polym. Prepr.,* 20 (1) (April 1979), pp. 956–959.

192. Changes in the dynamic mechanical properties of nylon-6 during ultraviolet light irradiation; **S. Yano and M. Murayama,** *ACS Polym. Prepr.,* 20 (1) (April 1979), pp. 964–966.

193. Dynamic mechanical properties of normal and burned human skin; **A. S. Kenyon and McC. Rayford,** *ACS Polym. Prepr.,* 20 (1) (April 1979), pp. 967–970.

194. Composition vs. mechanical properties of UV cured epoxy acrylate coatings for optical glass fibers; **H. N. Vazirani and T. K. Kwei,** *ACS Org. Coat. Plast. Chem.,* 40 (April 1979), p. 104–109.

195. Waterborne polymers for aircraft coatings; **L. W. Hill and D. E. Prince,** *ACS Org. Coat. Plast. Chem.,* 40 (April 1979), pp. 479–484.

196. Simultaneous interpenetrating networks based on castor oil elastomers and polystyrene synthesis and behavior; **N. Devia, J. A. Manson, L. H. Sperling, and A. Conde,** *ACS Org. Coat. Plast. Chem.,* 40 (April 1979), pp. 572–577.

197. The effects of moisture and stoichiometry on the dynamic mechanical properties of carbon reinforced epoxy composites; **J. D. Keenan, J. C. Seferis, and J. T. Quinlivan,** *ACS Org. Coat. Plast. Chem.,* 40 (April 1979), pp. 700–706.

198. Poly(2,6-dimethyl-1,4-phenylene oxide)-polystyrene interpenetrating polymer networks; **H. L. Frisch, K. C. Frisch, D. Klempner, and H. K. Yoon,** *ACS Org. Coat. Plast. Chem.,* 40 (April 1979), pp. 763–768.

199. Polyurethane-polyacrylate pseudo-interpenetrating polymer networks; **H. K. Yoon, D. Klempner, K. C. Frisch, and H. L. Frisch,** *ACS Org. Coat. Plast. Chem.,* 40 (April 1979), pp. 769–774.

200. Experimental analysis of hydrothermal aging in fiber reinforced epoxies; **R. M. Panos, R. P. Haak, P. J. Dynes, and D. H. Kaelble,** *ACS Org. Coat. Plast. Chem.,* 40 (April 1979), pp. 941–946.

201. Effects of extrusion on the structure and properties of high-impact polystyrene; **N. K. Kalfoglou and C. E. Chaffey,** *Polym. Eng. Sci.,* 19 (8) (June 1979), pp. 552–557.

202. Time-temperature cure behavior of epoxy based structural adhesives; **M. J. Doyle, A. F. Lewis, and H.-M. Li,** *Polym. Eng. Sci.,* 19 (10) (Aug. 1979), pp. 687–691.

203. Dynamic thermomechanical investigation of polymeric systems supported on inert substrates; **J. M. G. Cowie,** *Polym. Eng. Sci.,* 19 (10) (Aug. 1979), pp. 709–715.

204. Simultaneous interpenetrating networks based on castor oil elastomers and polystyrene. III. Morphology and glass transition behavior; **N. Devia, J. A. Manson, L. H. Sperling, and A. Conde,** *Polym. Eng. Sci.,* 19 (12) (Sept. 1979), pp. 869–877.

205. Linear and star branched butadiene-isoprene block copolymers and their hydrogenated derivatives; **J. E. McGrath, I. Wang, M. K. Martin, and K. S. Crane,** *ACS Polym. Prepr.,* 20 (2) (Sept. 1979), pp. 524–527.

206. Initial structure-property studies on a new series of segmented polyether-polyester copolymers; **P. C. Mody, G. L. Wilkes, D. A. Johnson, and K. B. Wegener,** *ACS Polym. Prepr.,* 20 (2) (Sept. 1979), pp. 539–542.

207. Characterization of a new binder coating; **S. L. Hager, J. E. Johnston, S. L. Watson, and T. B. MacRury,** *ACS Org. Coat. Plast. Chem.,* 41 (Sept. 1979), pp. 259–264.

208. Electron beam cured prepolymer based on polycaprolactone diol, isophoroned-iisocyanate and 2-hydroxyethylacrylate. I. Crosslink density and mechanical properties; **K. Park and G. L. Wilkes,** *ACS Org. Coat. Plast. Chem.,* 41 (Sept. 1979), pp. 308–312.

209. Study of crosslinking in polyimides by viscoelastic and diffusion techniques; **R. P. Chartoff and T. W. Chiu,** *Polym. Eng. Sci.,* 20 (4) (Mid-March 1980), pp. 244–251.

210. Crosslinking and mechanical properties of liquid rubber. I. Curative effect of aliphatic diols; **Y. Minoura, S. Yamashita, H. Okamoto, T. Matsuo, M. Izawa, and S.-I. Kohmotos,** *ACS Rubber Chem. Techn.,* 52 (5) (Nov.-Dec. 1979), pp. 920–948.

211. Low temperature dynamic properties of bitumin-rubber mixtures; **J. Shim-Ton, K. A. Kennedy, M. R. Piggott, and R. T. Woodhams,** *ACS Rubber Chem. Technol.,* 53 (1) (Mar.-Apr. 1980), pp. 88–106.

212. The influence of molecular geometry on the mechanical properties of homopolymers and block polymers of hydrogenated butadiene and isoprene; **Y. Mohajer, G. L. Wilkes, M. Martin, I. C. Wang, and J. E. McGrath,** *ACS Polym. Prepr.,* 21 (1) (March 1980), pp. 43–46.

213. Structural and mechanical properties of polybutadiene-containing polyurethanes; **C. M. Brunette, S. L. Hsu, W. J. MacKnight, and N. S. Schneider,** *ACS Polym. Prepr.,* 21 (1) (March 1980), pp. 181–182.

214. Thermal crosslinking of a chemically-modified ionomer; **M. J. Covitch, S. R. Lowry, C. L. Gray, and B. Blackford,** *ACS Polym. Prepr.,* 21 (2) (Aug. 1980), pp. 120–121.

215. Properties of some diblock copolymers based on 1,3 butadiene monomer; **R. E. Cohen, J. M. Torradas, D. E. Wilfong, and A. F. Halasa,** *ACS Polym. Prepr.,* 21 (1) (March 1980), pp. 216–217.

216. Compatibility and physical properties of a new polyester blend; **E. L. Slagowski, E. P. Chang, and J. J. Tkacik,** *ACS Polym. Prepr.,* 21 (1) (March 1980), pp. 278–279.

217. Oriented monofilaments from blends of poly(ethylene terephthalate) and polypropylene; **A. Rudin, D. A. Loucks, and J. M. Goldwasser,** *Polym. Eng. Sci.,* 20 (11) (July 1980), pp 741–746.

218. The degradation of an aromatic polyamideimide in an NO_2 atmosphere; **H. Kambe and R. Yokota,** *Polym. Eng. Sci.,* 20 (10) (Mid-July 1980), pp. 696–702.

219. Texture of meat by mechanical properties and microscopy; **A. S. Kenyon, J. D. Fairing, and McC. Rayford,** *J Food Sci.,* presented at 39th Annu. Meet. Inst. of Food Technologists, June 11, 1979.

220. Rheological characterization of polycaprolactam anionically synthesized in the presence of lithium chloride; **D. Acierno, R. D'Amico, F. P. LaMantia, and S. Russo,** *Polym. Eng. Sci.,* 20 (12) (Aug. 1980), pp. 783–786.

221. Polymer compatibility: nylon-epoxy resin blends; **Y. Y. Wang and S. A. Chen,** *Polym. Eng. Sci.,* 20 (12) (Aug. 1980), pp. 823–829.

222. Analysis of plasticizer activity in cellulose esters; **R. W. Seymour and J. R. Minter,** *Polym. Eng. Sci.,* 20 (18) (Dec. 1980), pp. 1188–1191.

223. Dynamic mechanical properties of high nitrile resins; **R. W. Yanik and A. D. McMaster,** *Polym. Eng. Sci.,* 20 (18) (Dec. 1980), pp. 1205–1213.

224. Morphology and mechanical properties of radiation polymerized urethane-acrylates; **L. H. Wadhwa and W. K. Walsh,** *ACS Div. Org. Coat. Plast. Chem.,* 42 (Mar. 1980), pp. 509–515.

225. Properties of blends of polyphenyl sulfone and reactive plasticizer; **S. Sikka and I. J. Goldfarb,** *ACS Div. Org. Coat. Plast. Chem.,* 43 (Aug. 1980), pp. 1–6.

226. New interpenetrating polymer networks based on industrial-type natural oils; **A. Qureshi, J. A. Manson, and L. H. Sperling,** *ACS Div. Org. Coat. Plast. Chem.,* 43 (Aug. 1980), pp. 7–12.

227. Non-uniform emulsion polymers. I. Process description and polymer properties; **D. R. Bassett and K. L. Hoy,** *ACS Div. Org. Coat Plast. Chem.,* 43 (Aug. 1980), pp. 611–615.

228. Changes in the dynamic modulus during thermal degradation of polyisoprene vulcanizates; **S. Yano,** *Rubber Chem. Technol.,* 53 (4) (Sept.-Oct. 1980), pp. 994–949.

229. Structure-property studies on a series of polycarbonate-polydimethylsiloxane block copolymers; **S. H. Tang, E. A. Meinecke, J. S. Riffle, and J. E. McGrath,** *Rubber Chem. Technol.,* 53 (5) (Nov.-Dec. 1980), pp. 1160–1169.

230. A critical analysis of the use of sample-supportive techniques in the measurement of dynamic mechanical relaxation processes; **R. M. Neumann, G. A. Senich, and W. J. MacKnight,** *Polym. Eng. Sci.,* 18 (8) (June 1978), pp. 624–627.

231. Polymer blends containing poly(vinylidene fluoride). I. Poly(alkyl acrylates); **D. C. Wahrmund, R. E. Bernstein J. W. Barlow, and D. R. Paul,** *Polym. Eng. Sci.,* 18 (9) (July 1978), pp. 677–682.

232. Polymer blends containing poly(vinylidene fluoride). II. Poly(vinyl esters); **R. E. Bernstein, D. R. Paul, and J. W. Barlow,** *Polym. Eng. Sci.,* 18 (9) (July 1978), pp. 683–686.

233. The operation of the rheovibron in the shear mode; **G. Locati,** *Polym. Eng. Sci.,* 18 (10) (Mid-Aug. 1981), pp. 793–798.

234. Polymer blends containing poly (vinylidene fluoride). III. Polymers containing ester, ketone, or ether groups; **R. E. Bernstein, D. C. Wahrmund, J. W. Barlow, and D. R. Paul,** *Polym. Eng. Sci.,* 18 (16) (Dec. 1978), pp. 1220–1224.

235. Error analysis and modelling of non-linear stress-strain behavior in measuring dynamic mechanical properties of polymers with the rheovibron; **A. R. Wedgewood and J. C. Seferis,** *Polymer,* 22 (Jul. 1981), pp. 966–991.

236. Phase transformation and ferroelectric polarization reversal in poly(vinylidene fluoride); **K. Matsushige and T. Takemura,** *ACS Div. Org. Coat. Plast. Chem.,* 44 (Mar.-Apr. 1981), pp. 250–256.

237. Rim urethanes structure/property relationships for linear polymers; **R. J. Zdrahala and F. E. Critchfield,** *ACS Div. Org. Coat. Plast. Chem.,* 44 (Mar.-Apr. 1981), pp. 275–279.

238. Introductory remarks on automated dynamic mechanical testing; **R. F. Boyer,** *ACS Div. Org. Coat. Plast. Chem.,* 44 (Mar.-Apr. 1981), pp. 492–502.

239. Dynamic mechanical spectroscopy using the autovibron (DDV-III-C); **S. M. Webler, J. A. Manson, and R. Lang** *ACS Div. Org. Coat. Plast. Chem.,* 44 (Mar.-Apr. 1981), pp. 524–529. (Also refer to *ACS Polym. Prepr.,* 22 (1) (Mar. 1981), p. 257.)

240. Morphology and properties of styrene and dimethylsiloxane triblock and multiblock copolymers; **S. K. Varshney and C. L. Beatty,** *ACS Div. Org. Coat. Plast. Chem.,* 44 (Mar.-Apr. 1981), pp. 698–704.

241. Thermoplastic interpenetrating polymer networks of a triblock copolymer elastomer and an ionomeric plastic. II. Mechanical behavior; **D. L. Siegfried, D. A. Thomas, and L. H. Sperling,** *Polym. Eng. Sci.,* 21 (1) (Jan. 1981), pp. 39–46.

242. Polymer compatibility: ternary blends of poly(vinylidene chloride-co-vinyl chloride), poly(vinyl chloride) and poly (acrylonitrile-co-butadiene); **Y.-Y. Wang and S.-A. Chen,** *Polym. Eng. Sci.,* 21 (1) (Jan. 1981), pp. 47–52.

243. Compatibility and physical properties of a new polyester blend; **E. L. Slagowski, E. P. Chang, and J. J. Tkacik,** *Polym. Eng. Sci.,* 21 (9) (Jun. 1981), pp. 513–517.

244. The physical aging of isotactic polypropylene; **M. K. Agarwal and J. M. Schultz,** *Polym. Eng. Sci.,* 21 (12) (Aug. 1981), pp. 776–781.

245. Thermal and mechanical properties of linear segmented polyurethanes with butadiene soft segments; **C. M. Brunette, S. L. Hsu, M. Rossmann, W. J. MacKnight, and N. S. Schneider,** *Polym. Eng. Sci.,* 21 (11) (Mid-Aug. 1981), pp. 668–674.

246. Polymer-performance on the dimensional stability and the mechanical properties of wood-polymer composites evaluated by polymer-wood interaction modes; **T. Hanada, S. Yoshizawa, I. Seo, and Y. Hashizume,** *ACS Div. Org. Coat. Plast. Chem.,* 45 (Aug. 1981), pp. 375–381.

247. New polymers from botanical oils; **S. Qureshi, J. A. Manson, L. H. Sperling, and C. J. Murphy,** *ACS Div. Org. Coat. Plast. Chem.,* 45 (Aug. 1981), pp. 649–654.

248. Morphology and mechanical properties of filled polyethylene; **V. P. Chacko, R. J. Farris, and F. E. Karasz,** *ACS Div. Org. Coat. Plast. Chem.,* pp. 767–771.

249. Morphology and properties of styrene and dimethylsiloxane triblock and multiblock copolymers; **S. K. Varshney, C. L. Beatty, and P. Bajaj,** *ACS Polym. Prepr.,* 22 (1) (Mar. 1981), pp. 321–323.

250. bis-methacryloxy bisphenol-A epoxy networks; synthesis, characterization, thermal and mechanical properties; **A. K. Banthia, I. Yilgor, J. E. McGrath, and G. Wilkes,** *ACS Polym. Prepr.,* 22 (1) (Mar. 1981), pp. 209–211.

251. Relaxations in crosslinked aromatic backbone polymers; **J. A. Hinkley and F. J. Campbell,** *ACS Polym. Prepr.,* 22 (2) (Aug. 1981), pp. 297–298.

252. Dynamic properties of synthetic elastomers and the Schwarzl relationship; **A. E. Hirsch,** *ACS Polym. Prepr.,* 22 (2) (Aug. 1981), p. 195.

253. Relation between dynamic mechanical properties and dye diffusion behavior in acrylic fibers from polymer blends; **Y. G. Bryant and T. Murayama,** *J. Appl. Polym. Sci.,* 24 (1979), pp. 2389–2397.

254. Dynamic compression mechanical properties of fiber masses with a rheovibron viscoelastometer; **T. Murayama,** *J. Appl. Polym. Sci.,* 25 (1980), pp. 529–534.

255. Reinforcement of acrylonitrile copolymer by polymer blending; **E. L. Lawton, T. Murayama, V. F. Holland, and D. C. Felty,** *J. Appl. Polym. Sci.,* 25 (1980), pp. 187–209.

256. Evaluation of fatigue lifetime and elucidation of fatigue mechanism in plasticized poly(vinyl chloride) in terms of dynamic viscoelasticity; **A. Takahara, T. Kajiyama, M. Takayanagi, and K. Yamada,** *J. Appl. Polym. Sci.,* 25 (1980), pp. 597–614.

257. Application of the rheovibron to inorganic glass problems. I. The mixed alkali effect loss spectrum; **T. Atake and C. A. Angell,** *J. Non-Cryst. Solids,* 38 and 39 (1980), pp. 439–444.

258. Haloaldehyde polymers. XXIII. Thermal and mechanical properties of chloral polymers; **P. Kubisa, L. S. Corley, T. Kondo, M. Jacovic, and O. Vogl,** *Polym. Eng. Sci.,* 21 (13) (Sept. 1981), pp. 829–838.

259. Properties of polyether-polyurethane zwitterionomers; **K. K. S. Hwang, C.-Z. Yang, and S. L. Cooper,** *Polym. Eng. Sci.,* 21 (15) (Oct. 1981), pp. 1027–1036.

260. Mechanical and thermal properties of aromatic copolyesters containing rigid binaphthyl units; **J. Manusz, R. W. Lenz, and W. J. MacKnight,** *Polym. Eng. Sci.,* 21 (16) (Nov. 1981), pp. 1079–1084.

261. Poly(*n*-butyl acrylate)/polystyrene interpenetrating polymer networks and related materials. I. Dynamic mechanical spectroscopy and morphology; **J. K. Yeo, L. H. Sperling, and D. A. Thomas,** *Polym. Eng. Sci.,* 22 (3) (Feb. 1982), pp. 190–196.

262. Dynamic mechanical study of the curing reaction of an unsaturated polyester; **C. Y. Yap and H. L. Williams,** *Polym. Eng. Sci.,* 22 (4) (Mar. 1982), pp. 254–259.

263. The mechanical properties of styrene-butadiene-styrene (SBS) triblock copolymer blends with polystyrene (PS) and styrene-butadiene copolymer (SBR); **J. Diamant, D. Soong, and M. C. Williams,** *Polym. Eng. Sci.,* 22 (11) (Mid-Aug. 1982), pp. 673–683.

264. Studies of poly(2,6-dimethyl-1,4-phenylene oxide) blends. I. Copolymers of styrene and maleic anhydride; **J. R. Fried and G. A. Hanna**, *Polym. Eng. Sci.*, 22 (11) (Mid-Aug. 1982), pp. 705–718.

265. Determination of the glass transition of polymer by the autovibron; **T. Murayama**, *Polym. Eng. Sci.*, 22 (12) (Mid-Aug. 1982), pp. 788–792.

266. Effect of sub-T_g relaxations on the gas transport properties of polyesters; **R. R. Light and R. W. Seymour**, *Polym. Eng. Sci.*, 22 (14) (Mid-Oct. 1982), pp. 857–864.

267. Dynamic mechancial behavior of filled polyethylenes and model composites; **V. P. Chacko, F. E. Karasz, and R. J. Farris**, *Polym. Eng. Sci.*, 22 (15) (Oct. 1982), pp. 968–974.

268. Effect of short glass fibers and particulate fillers on fatigue crack propagation in polyamides; **R. W. Lang, J. A. Manson, and R. W. Hertzberg**, *Polym. Eng. Sci.*, 22 (15) (Oct. 1982), pp. 982–987.

269. Recent advances in interpenetrating polymer networks; **K. C. Frisch, D. Klempner, and H. L. Frisch**, *Polym. Eng. Sci.*, 22 (17) (Mid-Dec. 1982), pp. 1143–1152.

270. Thermal and dynamic mechanical analysis of polycarbonate/poly(methyl methacrylate) blends; **Z. G. Gardlund**, *ACS Polym. Prepr.*, 23 (1) (Marc. 1982), pp. 258–259.

271. Polytetramethylene ether glycol: effect of concentration, molecular weight, and molecular weight distribution on properties of MDI/BDO-based polyurethanes; **E. Pechhold and G. Pruckmayr**, *Rubber Chem. Technol.*, 55 (1) (Mar.-Apr. 1982), pp. 76–87.

272. Morphology-property relationships in EPDM-polybutadiene blends; **G. R. Hamed**, *Rubber Chem. Technol.*, 55 (1) (Mar.-Apr. 1982), pp. 151–160.

273. Thermal transition and relaxation behavior of polybutadiene polyurethanes based on 2,6-toluene diisocyanate; **C. M. Brunette and W. J. MacKnight**, *Rubber Chem. Technol.*, 55 (5) (Nov.-Dec. 1982), pp. 1413–1425.

274. Acetylene terminated resin mechanical characterization. I. Dynamic mechanical properties and fracture energies at different cure states; **C. L. Leung**, *ACS Div. Org. Coat. Appl. Polym. Sci. Proc.*, 46 (Mar. 1982), pp. 322–327.

275. Ageing of structural film adhesives — changes in chemical and physical properties and the effect on joint strength; **C. E. M. Morris, P. J. Pearce, and R. G. Davidson**, *ACS Div. Org. Coat. Appl. Polym. Sci. Proc.*, 47 (Sept. 1982), pp. 79–83.

276. Dynamic mechanical and dielectric properties of an epoxy resin during cure; **W. X. Zukas and W. J. MacKnight**, *ACS Div. Org. Coat. Appl. Polym. Sci. Proc.*, 47 (Sept. 1982), pp. 425–428.

277. Nondestructive evaluation of some bonded joints; **T. C. Ward, M. Sheridan, and D. L. Kotzev**, *ACS Div. Org. Coat. Appl. Polym. Sci. Proc.*, 47 (Sept. 1982), pp. 467–470.

278. Measurement of dynamic viscoelastic properties of polymer melts and liquids by the modified rheovibron; **T. Murayama**, *J. Appl. Polym. Sci.*, 27 (1982), pp. 89–96.

279. Measurement of dynamic transverse compressional properties of hollow fibres; **T. Murayama**, *Polym. Eng. Rev.*, 2 (1) (1982), pp. 84–96. (Also refer to *Proceedings of the 1982 Joint Conference on Experimental Mechanics*, II (May 1982), pp. 1007–1013.)

280. Study of the kinetics of *in situ* polymerization in wood by dynamic mechanical measurements; **R. V. Subramanian and R. Hoffmann**, *J. Polym. Sci. Polym. Lett. Ed.*, 20 (1982), pp. 001–005.

281. Dynamic infrared linear dichroism of polymer films under oscillatory deformation; **I. Noda, A. E. Dowrey, and C. Marcott**, *J. Polym. Sci. Polym. Lett. Ed.*, 21 (1983), pp. 99–103.

282. Batch and semicontinuous emulsion copolymerization of vinyl acetate-butyl acrylate. II. Morphological and mechanical properties of copolymer latex films; **S. C. Misra, C. Pichot, M. S. El-Aasser, and J. W. Vanderhoff**, *J. Polym. Sci., Polym. Chem. Ed.*, 21 (8) (Aug. 1983), pp. 2383–2396.

283. Anisotropic electrical conductivity of drawn elastomeric ionene-TCNQ salts; **M. Watanabe, Y. Takizawa, and I. Shinohara**, *J. Polym. Sci., Polym. Chem. Ed.*, 21 (8) (Aug. 1983), pp. 2397–2404.

284. Fast ion motion in glassy and amorphous materials; **C. A. Angell,** *Solid State Ionics,* 9–10 (1983), pp. 3–16.

285. Structure-properties relationships in biaxially oriented polypropylene films; **A. J. DeVries,** *Polym. Eng. Sci.,* 23 (5) (Mid-Apr. 1983), pp. 241–246.

286. Properties of polyisobutylene-polyurethane block copolymers. I. Macroglycols from ozonolysis of isobutylene-isoprene copolymer; **T. A. Speckhard, G. Ver Strate, P. E. Gibson, and S. L. Cooper,** *Polym. Eng. Sci.,* 23 (6) (Apr. 1983), pp. 337–349.

287. Dynamic mechanical and thermomechanical properties of silicone-polycarbonate block copolymers; **W. Maung, K. M. Chua, T. H. Ng, and H. L. Williams,** *Polym. Eng. Sci.,* 23 (8) (Mid-Jun. 1983), pp. 439–445.

288. Blending of polystyrene with styrene-siloxane block and graft copolymers; **P. Bajaj, D. C. Gupta, and S. K. Varshney,** *Polym. Eng. Sci.,* 23 (15) (Oct. 1983), pp. 820–825.

289. Mechanical behavior of some epoxies with epoxidized natural oils as reactive diluents; **S. Qureshi, J. A. Manson, R. W. Hertzberg, and L. H. Sperling;** *ACS Div. Org. Coat. Appl. Polym. Sci. Proc.,* 48 (Mar. 1983), pp. 576–580.

290. Fatigue crack propagation in short-glass-fiber-reinforced styrene-acrylonitrile copolymer; **R. W. Lang, J. A. Manson, and R. W. Hertzberg,** *ACS Div. Org. Coat. Appl. Polym. Sci. Proc.,* 48 (Mar. 1983), pp. 816–821.

291. All aromatic biphenylene end-capped high performance composite resins; **J. P. Droske and J. K. Stille,** *ACS Div. Org. Coat. Appl. Polym. Sci. Proc.,* 48 (Mar. 1983), pp. 924–928.

292. Polycarbonate-poly(methyl methacrylate) block copolymers: effect of composition on transition temperatures; **Z. G. Gardlund,** *ACS Polym. Prepr.,* 24 (2) (Aug. 1983), pp. 18–19.

293. Solid state properties of segmented polysiloxane thermoplastic elastomeric copolymers; **D. Tyagi, I. Yilgor, G. L. Wilkes, and J. E. McGrath,** *ACS Polym. Prepr.,* 24 (2) (Aug. 1983), pp. 39–40.

294. Effect of molecular orientation on bonded joints; **T. C. Ward, A. W. Brinkley, M. Sheridan, and P. Koning,** *ACS Polym. Prepr.,* 24 (2) (Aug. 1983), pp. 119–121.

295. Thermal, mechanical and spectroscopic studies of linear segmented polyurethane blends. II. Blends of polyester urethanes with a styrene/acrylonitrile copolymer; **M. Iskander, C. Tran, and J. E. McGrath,** *ACS Polym. Prepr.,* 24 (2) (Aug. 1983), pp. 126–129.

296. Ionic crosslinking of carboxylated SBR; **K. Sato,** *Rubber Chem. Technol.,* 56 (5) (Nov.-Dec. 1983), pp. 942–958.

297. Structure-property relationships in isocyanurate-based urethane elastomers; **C. L. Wang, D. Klempner, and K. C. Frisch,** *ACS Polym. Mater. Sci. Eng.,* 49 (Aug. 1983), pp. 38–44.

298. Properties of model polyether polyurethane block copolymers; The effect of hard segment length and polydispersity; **S. B. Lin, K. K. S. Hwang, K. S. Wu, S. Y. Tsay, and S. L. Cooper,** *ACS Polym. Mater. Sci. Eng.,* 49 (Aug. 1983), pp. 53–57.

299. A further study of quantitative evaluation of the fine amorphous structure distribution of a semi-crystalline polymer solid from the dynamic loss tangent (tan δ) vs. temperature curve; **S. Manabe and K. Kamide,** *Polym. J.,* 16 (5) (1984), pp. 375–389.

300. The effect of variations in chain connectivity on strain-induced crystallization in unvulcanized EPDM elastomers; **B. J. R. Scholtens,** *Rubber Chem. Technol.,* 57 (4) (Sept.-Oct. 1984), pp. 703–724.

301. A new drawing technique for nylon 6 by reversible plasticization with iodine; **H. H. Chuah and R. S. Porter,** *ACS Polym. Mater. Sci. Eng.,* 51 (1984), pp. 325–328.

302. Effects of crystallinity on the microdomain morphology of short segment block copolymers; **J. L. Castles, M. A. Vallance, S. L. Cooper, and J. M. McKenna,** *ACS Polym. Mater. Sci. Eng.,* 51 (1984), pp. 387–391.

303. Recent studies on interpenetrating polymer networks at the Polymer Institute, University of Detroit; **D. Klempner, X. H. Xiao, K. C. Frisch, E. Cassidy, and H. L. Frisch,** *ACS Polym. Mater. Sci. Eng.,* 51 (1984), pp. 503–511.

304. Mechanical properties and molecular aggregation of soluble aromatic polyimide; **M.Kochi, R. Yokota, and I. Mita,** *Polym. Eng. Sci.,* 24 (13) (Sept. 1984), pp. 1021–1025.

305. Surface-modified kevlar fiber-reinforced ionomer: dynamic mechanical properties and their interpretation; **M. Takayanagi and T. Katayose,** *Polym. Eng. Sci.,* 24 (13) (Sept. 1984), pp. 1047–1050.

306. Microphase structure and properties of polyurethane/polyvinyl interstitial composites; **J. T. Koberstein and R. S. Stein,** *Polym. Eng. Sci.,* 24 (5) (Mid-Apr. 1984), pp. 293–310.

307. Network structure in bis-phthalonitrile polymers; **J. A. Hinkley,** *ACS Polym. Prepr.,* 25 (1) (Apr. 1984), pp. 102–103.

308. Some novel diene polymers prepared with lanthanide catalysts; **H. C. Yeh and H. L. Hsieh,** *ACS Polym. Prepr.,* 25 (2) (Aug. 1984), pp. 52–53.

309. Viscoelastic properties of radial substituted styrene-isoprene block copolymers; **M. M. Sheridan, J. M. Hoover, T. C. Ward, and J. E. McGrath,** *ACS Polym. Prepr.,* 25 (2) (Aug. 1984), pp. 102–104.

310. Dynamic mechanical properties of some polyurethane-acrylic copolymer interpenetrating polymer networks; **R. B. Fox, J. L. Bitner, J. A. Hinkley, and W. Carter,** *Polym. Eng. Sci.,* 25 (3) (Feb. 1985), pp. 157–163.

311. Evidence of dual mobility in sulfur dioxide-polyarylate system: an integral sorption analysis; **G. R. Ranade, R. Chandler, C. A. Plank, and W. L. S. Laukhuf,** *Polym. Eng. Sci.,* 25 (3) (Feb. 1985), pp. 164–169.

312. Damping modulus studies of poly(dimethylsiloxane)-bisphenol-A-polycarbonate block copolymers; **W. Maung & H. L. Williams,** *Polym. Eng. Sci.,* 25 (2) (Mid-Feb. 1985), pp. 113–117.

313. Polymerization of butadiene and isoprene with lanthanide catalysts; characterization and properties of homopolymers and copolymers; **H. L. Hsieh and H. C. Yeh,** *Rubber Chem. Technol.,* 58 (1) (Mar.-Apr. 1985), pp. 117–145.

314. Dynamic mechancial analysis of sulfonated polystyrene ionomers; **R. A. Weiss,** *ACS Polym. Prepr.,* 26 (1) (Apr. 1985), pp. 21–22.

315. Some observations regarding curing of free radical network resins and their respective mechanical behavior; **B. Orler, G. L. Wilkes, and N. Ganesh Kumar,** *ACS Polym. Prepr.,* 26 (1) (Apr. 1985), pp. 281–283.

316. Rates of property change in plasticized filled PVC compounds; **A. Ajji, A. Rudin, and H. P. Schreiber,** *ACS Polym. Prepr.,* 26 (1) (Apr. 1985), pp. 306–307.

317. Dynamic mechanical properties of ultradrawn high molecular weight isotactic polypropylene; **S. J. Roy and R. St. John Manley,** *ACS Polym. Mater. Sci. Eng.,* 52 (Mar. 1985), pp. 53–56.

318. Interpenetrating polymer networks for vibration attenuation; **D. Klempner, L. Berkowski, K. C. Frisch, K. H. Hsieh, and R. Ting,** *ACS Polym. Mater. Sci. Eng.,* 52 (Mar. 1985), pp. 57–63.

319. Viscoelasticity of calf hide impregnated with radiation-polymerized hydrogel; **P. L. Kronick, R. Artymyshyn, P. R. Buechler, and W. Wise,** *ACS Polym. Mater. Sci. Eng.,* 52 (Mar. 1985), pp. 186–191.

320. Effect of particle's morphology on grafting reactions in core/shell latexes; **M. P. Merkel, V. L. Dimonie, M. S. El-Aasser, and J. W. Vanderhoff,** *ACS Polym. Mater. Sci. Eng.,* 52 (Mar. 1985), pp. 320–324.

321. Properties of coatings containing non-reactive microgels; **A. Kashihara, K. Ishii, K. Kida, S. Ishikura, and R. Mizuguchi,** *ACS Polym. Mater. Sci. Eng.,* 52 (Mar. 1985), pp. 453–457.

Chapter 2

CONVENTIONAL VISCOMETERS AND GENERAL CONCEPTS

I. INTRODUCTION

A variety of viscometers are used in the polymers and plastics industries. These devices fall into two general categories: scientific instruments designed to make basic measurements of rheological properties and instruments that are primarily employed for industrial quality control. This chapter outlines the operating principles and basis for both general classes of instruments. It should be noted that a wide variety of rheological test methods and instruments are used in other industries. A summary of the major rheological measurement techniques for different industries is given in the last section of this chapter.

II. ANALYSIS OF CAPILLARY FLOW

The capillary tube viscometer is one of the simplest and most widely used instruments for rheological characterization. Before describing its operation one must first consider the problem of capillary flow, as in Figure 1, in which a fluid is forced at a steady rate from a large reservoir into a small diameter capillary tube of length L. Assume steady, isothermal flow of an incompressible fluid and $L/R \gg 1$. The coordinate system is defined along the r, θ, and z axes. If entrance and exit losses in the capillary tube are ignored, then

$$V_\theta = 0 \quad ; \quad V_r = 0$$

$$V_z \neq 0 \quad ; \quad V_z \neq f(\theta)(\text{axisymmetry}) \tag{1}$$

Writing the continuity equation

$$\frac{\partial P}{\partial t} + \frac{1}{r} \frac{\partial}{\partial r} (\rho r V_r) + \frac{1}{r} \frac{\partial}{\partial \theta} (\rho V_\theta) + \frac{\partial}{\partial z} (\rho V_z) = 0 \tag{2}$$

where density ρ is assumed to be constant.

Because the system is at steady state $(\partial P/\partial t) = 0$, since $V_r = 0$, $V_\theta = 0$, the second and third terms on the left side of Equation 2 drop out as well, leaving:

$$\frac{\partial V_z}{\partial z} = 0 \tag{3}$$

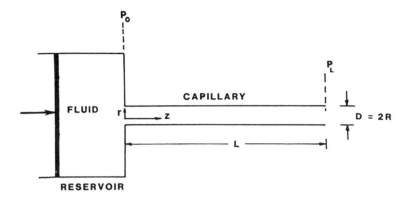

FIGURE 1.   Capillary tube flow.

The momentum equation for the z-component is

$$P\left[\frac{\partial V_z}{\partial t} + V_r \frac{\partial V_z}{\partial r} + \frac{V_\theta}{r}\frac{\partial V_z}{\partial \theta} + V_z \frac{\partial V_z}{\partial z}\right] = -\frac{\partial P}{\partial z}$$

$$-\left[\frac{1}{r}\frac{\partial}{\partial r}(r\,\tau_{rz}) + \frac{1}{r}\frac{\partial \tau_{\theta r}}{\partial \theta} + \frac{\partial \tau_{zz}}{\partial z}\right] + \rho g_z \qquad (4)$$

From the constraints of the system imposed, the z-component momentum equation reduces to

$$\frac{dP}{dz} = -\frac{1}{r}\frac{d}{dr}(r\,\tau_{rz}) \qquad (5)$$

Both sides of this expression are a function of r [i.e., f(r)] and both sides are constant. It can, therefore, be assumed that dP/dz $\simeq$ ΔP/L (note that dP/dz $<$ 0), and can therefore integrate Equation 5:

$$r\,\tau_{rz} = -\frac{\Delta P}{L}r^2 + C_1$$

$$\tau_{rz} = -\frac{\Delta P}{L}r + \frac{C_1}{r} \qquad (6)$$

At r = 0, $\tau_{zr} \neq \infty$ therefore, C_1 = 0 and hence:

$$\tau_{zr} = -\left(\frac{\Delta P}{2L}\right)r \qquad (7)$$

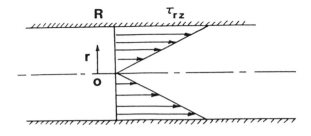

FIGURE 2. Shear stress profile in tube flow.

Equation 7 describes the shear stress profile for any incompressible fluid, and the profile across the capillary tube will have the form shown in Figure 2.

Now, if the rheological properties of the fluid follow that a power law constitutive equation, i.e., $\tau = -K|\dot\gamma|^{n-1}\dot\gamma$ where τ and τ_{zr}, $\dot\gamma = dV_z/dr$ and K and n are constants then

$$-K\left|\frac{dV_z}{dr}\right|^{n-1}\frac{dV_z}{dr} = -\left(\frac{\Delta P}{2L}\right)r \qquad (8)$$

Note that

$$\left|\frac{dV_z}{dr}\right| = -\frac{dV_z}{dr}$$

since $dV_z/dr < 0$.

This expression may now be integrated and the "no-slip" at the wall condition may be imposed

$$V_z(r) = \left(\frac{nR}{n+1}\right)\left[\frac{\Delta PR}{2KL}\right]^{1/n}\left[1 - \left(\frac{r}{R}\right)^{(n+1)/n}\right] \qquad (9)$$

Equation 9 is the velocity profile of a power law fluid in tube flow. For $n = 1$, $K = \mu$; i.e., the limit of a Newtonian fluid is reached and the velocity profile of Equation 9 reduces to the familiar parabolic velocity profile expression

$$V_z(r) = \frac{\Delta PR^2}{4\mu L}\left[1 - \left(\frac{r}{R}\right)^2\right] \qquad (10)$$

The theoretical velocity profiles computed from Equation 9 have the shapes shown in Figure 3: for $n = 1$ the parabolic shape (Newtonian, i.e.,

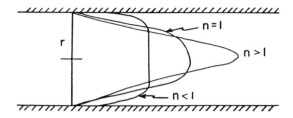

FIGURE 3. Velocity profiles generated by Equation 9: n = 1, Newtonian fluid; n > 1, dilatant fluid; n < 1, pseudoplastic fluid.

Equation 10); for n > 1 a dilatant fluid; for n < 1 a pseudoplastic fluid. For polymer melts and solutions, n < 1, and hence velocity profiles are typically "plug-like". This means that there are very high velocity gradients in the vicinity of the tube wall, as shown in Figure 3.

In the above analysis, the constitutive equation for the velocity profile of a power law fluid was derived from first principles. This implies that one may calculate the volumetric flow rate of the fluid through the tube knowing V_z (r).

$$Q = 2\pi \int_0^R V_z(r)\ r\ dr \tag{11}$$

Substituting in the velocity profile expression Equation 9 and integrating, gives

$$Q = \frac{\pi n R^3}{3n + 1} \left[\frac{R\Delta P}{2KL} \right]^{1/2} \tag{12}$$

Equation 12 can be written in a linear form as:

$$\ln Q = \ln\left(\frac{n\pi R^3}{3n + 1} \right) + \frac{1}{n} \ln\left(\frac{R}{2KL} \right) + \frac{1}{n} \ln \Delta P \tag{13}$$

The striking observation of Equation 13 is that the first two sets of expressions on the right side are constant. Hence, $Q = f(\Delta P)$ and a logarithmic plot of Q vs. ΔP will give a straight line. Nature and mathematics do not always agree. Actual flow data on pseudoplastic materials tend to be nonlinear, as shown in Figure 4, the reason being that $n = f\ (dV_z/dr)$; i.e., n tends to depend on Q. Although from a qualitative standpoint, the power law constitutive equation shows proper trends, quantitatively it is a poor predictive tool. From the standpoint of rheological characterization, there is no substitute for experiments and therefore, a capillary tube flow experiment must be designed that is general enough to evaluate all types of time-independent fluids.

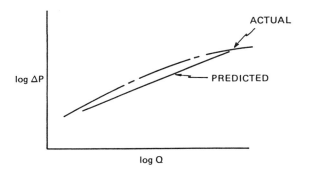

FIGURE 4. Comparison of actual and theoretical flow curves.

III. CAPILLARY VISCOMETRY

Orienting the capillary tube so that the flow is vertically downward, one may write the force balance

$$\frac{\pi D^2}{4} \Delta P = \pi DL \ \tau_w$$

or

$$(\tau_{rx})_{r=R} = \tau_w = \frac{D\Delta P}{4L} \tag{14}$$

Following the development of the Rabinowitsch-Mooney equation for steady, laminar flow of a time-independent fluid through a vertical tube, the author considered the flow in a slice of fluid constituting an annulus between r and r + dr:

$$dQ = V \ 2\pi r \ dr$$

where V is the linear velocity at r. Hence,

$$Q = \pi \int_0^R V \ 2r \ dr = \pi \int_0^{R^2} V \ d(r^2)$$

Applying the no-slip condition (V = 0, r = R), the expression becomes

$$\frac{Q}{\pi R^3} = \frac{8Q}{\pi D^3} = \frac{1}{\tau_w^3} \int_0^{\tau_w} \tau_{rx}^2 \ f(\tau_{rx}) d\tau_{rx} \tag{15}$$

where $f(\tau_{rx}) = -dV/dr$. By appropriate manipulation and replacing τ_w with $D\Delta P/4L$ we obtain the Rabinowitsch-Mooney equation of shear rate at the tube wall for steady, laminar flow of a time-independent fluid:

$$3\left(\frac{8Q}{\pi D^3}\right) + \frac{D\Delta P}{4L}\frac{d(8Q/\pi D^3)}{d(D\Delta P/4L)} = \left(-\frac{dV}{dr}\right)_w \qquad (16)$$

Since $Q = VA = \frac{1}{4}\pi D^2 V$, Equation 16 can be rearranged:

$$3\left(\frac{8Q}{\pi D^3}\right) = \left(-\frac{dV}{dr}\right)_w = \frac{3}{4}\left(\frac{8V}{D}\right) + \left(\frac{8V}{D}\right)\frac{d[\frac{1}{4}(8V/D)]/(8V/D)}{d(D\Delta P/4L)/(D\Delta P/4L)}$$

or

$$-\dot{\gamma}_w = \left(-\frac{dV}{dr}\right)_w = \frac{8V}{D}\left[\frac{3}{4} + \frac{1}{4n}\right] = \frac{3n+1}{4n}\left(\frac{8V}{D}\right) \qquad (17a)$$

where

$$n = \frac{d\,\ln(D\Delta P/L)}{d\,\ln(8V/D)}$$

This expression as may also be stated

$$-\dot{\gamma}_w = \frac{3}{4}\Gamma + \frac{\tau_w}{4}\frac{d\Gamma}{d\tau_w} \qquad (17b)$$

where $\Gamma = 4Q/\pi R^3$ is the shear rate for a Newtonian fluid.

The Poiseuille equation for a Newtonian fluid (Bird et al., 1967) is

$$\Delta P = \frac{8\mu LV}{R^2} = \frac{32\mu LV}{D^2}$$

which can also be written as

$$\ln\left(\frac{D\Delta P}{4L}\right) = \ln\left(\frac{8V}{D}\right) + \ln(\mu)$$

Because μ is constant,

$$\frac{d\,\ln(D\Delta P/4L)}{d\,\ln(8V/D)} = n = 1.0$$

Substituting this expression into Equation 17 shows that $(-dV/dr)_w = 8V/D$. This statement is true only for a Newtonian fluid because viscosity is a constant. In the case of a non-Newtonian n varies along the log-log plot of $D\Delta P/4L$ vs. $8V/D$. To apply a capillary tube viscometer in order to develop a non-Newtonian fluid's flow curve, data must be converted into a logarithmic plot of $D\Delta P/4L$ vs. $8V/D$, where n is determined as the slope of the curve at a particular value of the wall shear stress (i.e., $\tau_w = D\Delta P/4L$). The shear rate corresponding to this wall shear stress value is determined from Equation 17. It is important to note that Equation 17 is derived from laminar flow; hence, one must ensure that capillary tube flow is in this regime by checking the Reynolds number. Metzner and Reed (1955) have derived the generalized Reynolds number for a power law fluid as:

$$Re_{gen} = \frac{D^n V^{2-n} \rho}{\dot{\gamma}} \tag{18}$$

or

$$Re_{gen} = \frac{D^n V^{2-n} \rho}{K 8^{n-1}} \tag{19}$$

This expression applies to all fluids which are not time dependent. In fact, Metzner and Reed (1955) have shown that such non-Newtonian fluids obey the conventional Newtonian friction factor vs. Reynolds number. Therefore, laminar flow prevails for $Re_{gen} < 2100$.

Data generated in any capillary tube viscometer will require corrections for four contributions to frictional losses: (1) the head of fluid above the tube exit; (2) kinetic energy effects; (3) tube entrance effects; (4) effective slip near the tube wall. The last effect is not always present and generally can be detected as explained later.

The first three corrections can be made through application of the total mechanical energy balance. Skelland (1967) outlines a correction approach and provides an illustrative example. An alternative approach that generally gives good results is the Bagley (1954) shear stress correction method.

This method was originally developed using a variety of polyethylenes. The resulting flow curve is independent of the L/D ratio of the capillary; however, the procedure requires the assumptions of zero slip at the tube walls and time independence of the fluid material. The method uses a fictitious length of tubing $N_f R$ that is added to the actual length of the tube L, such that the measured total ΔP across L is that which would be obtained in fully developed flow over the length $(L + N_f R)$ at the flow rate used in the experiment. The shear stress at the wall for fully developed flow over the length $(L + N_f R)$ is then

$$\tau_w = R(\Delta P/2)(L + N_f R) \tag{20}$$

If the fluid is time independent and there is no slip at the tube wall, then τ_w is a unique function of $4Q/\pi R^3 (\simeq \Gamma)$ in laminar flow [see Skelland (1967) for derivation]. Hence

$$\tau_w = \frac{R}{2}\left(\frac{\Delta P}{L + N_f R}\right) = f\left(\frac{4Q}{\pi R^3}\right) \tag{21}$$

from which

$$\frac{L}{R} = -N_f + \frac{\Delta P}{2f(4Q/\pi R^3)} \tag{22}$$

The procedure involves obtaining a series of ΔP measurements made on several tubes of various L/R (or L/D) ratios, while maintaining $4Q/\pi R^3$ constant. Further discussions on this technique can be found in Chapter 3.

It is important to note that some non-Newtonian fluids display a peculiar orientation of their molecules in the vicinity of the tube walls. In an aqueous suspension, the discrete phase may actually move away from the wall, leaving a thin layer of the continuum phase in the immediate proximity of the wall. In this situation a reduction occurs in the apparent viscosity in the vicinity of the wall, known as the phenomenon of effective slip at the wall. For a capillary tube viscometer, an effective slip coefficient can be evaluated as a function τ_w using the following relationship:

$$\frac{Q}{\pi R^3 \tau_w} = \frac{\beta}{R} + \frac{1}{\tau^4}\int_0^{\tau_w} \tau_{rx}^2\, f(\tau_{rx}) d\,\tau_{rx} \tag{23}$$

A series of capillary tube measurements is needed in order to evaluate β. For a range of tubes of various R but constant L, a plot of $Q/R^3\tau_w$ vs. τ_w can be constructed. If no slip occurs with the material (i.e., $\beta = 0$), all the curves will coincide in accordance with Equation 23. If, however, a family of curves is obtained, then β has some value. From the family of curves, select some fixed value of τ_w and obtain values of $Q/\pi R^3 \tau_w$ for each corresponding R-value curve. Now prepare a plot of $Q/\pi R^3 \tau_w$ vs. $1/R$. The slope of this curve provides a β-value for the selected τ_w, according to Equation 23. By repeating this procedure for different τ_w-values a relationship between τ_w and β can be established. Figure 5 illustrates the method. Once a relationship between the effective slip coefficient and wall shear stress is established, the measured volumetric flow rate from the capillary tube can be corrected for slippage:

$$Q' = Q_{meas} - \beta\, \tau_w\, \pi R^2 \tag{24}$$

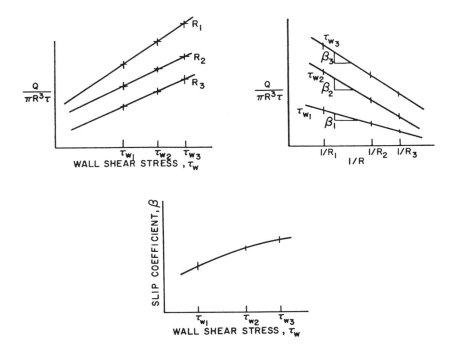

FIGURE 5. The graphic method for estimating for effective wall slippage.

where Q' is the volumetric flow rate corrected for slippage and $\tau_w = D\Delta P/4L$ corresponding to $Q_{measured}$.

IV. CONCENTRIC CYLINDER ROTARY AND ROTATING CYLINDER VISCOMETERS

Concentric cylinder rotary and rotating cylinder viscometers are most often applied to solution viscosity measurements and are usually limited to shear rates of <100 s^{-1}.

The basic elements of the concentric cylinder rotary viscometer are shown in Figure 6. The unit operates by applying shear to a fluid located in the annulus between the concentric cylinders. One cylinder (usually the cup) rotates while the other is fixed in space. From a series of measurements of the angular speed of the rotating cup and of the torque applied to the stationary cylinder, a flow curve for the fluid under shear can be derived.

The development of the constitutive equations for this type of viscometer is as follows. From a torque balance about the surface of the fixed bob while the cup revolves at a steady angular velocity, one may write:

$$\tau_b(2\pi R_b \ell)R_b = 2\pi R_b^2 \ell\, \tau_b \tag{25}$$

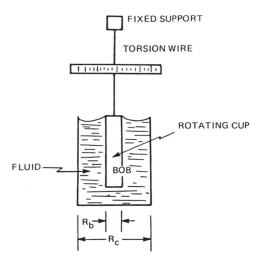

FIGURE 6.   Basic components of the concentric cylinder rotary viscometer.

where T = torque, R_b = bob radius, and τ_b = shearing stress at the bob surface.

The balance equation neglects end effects at the base of the bob and assumes that shear occurs only at the cylindrical surface of the bob. In practice, however, end effects are important and can be accounted for by including an effective bob length ℓ_{eff} in Equation 25 for the bob surface shear stress,

$$\tau_b = T/(2\pi R_b^2 \ell) \tag{26}$$

The shear rate at the bob surface is a complex relationship, and for brevity is not included here. An approximate relationship proposed by Calderbank and Moo-Young (1959) based on the analytical derivation of Kreiger and Maron (1954) is

$$\left(\frac{dV}{dr}\right)_b = \frac{4\pi N}{1 - (R_c/R_b)^2} C_R \tag{27}$$

Equation 27 is applicable for cup-to-bob radii ratios smaller than 1.75. The term C_R represents a series expansion of the shear rate-geometric relationship developed by Kreiger and Maron for time-independent power law fluids. Table 1 provides values of C_R that can be used in Equation 27. The rheological flow curve can be developed by a logarithmic plot of τ_b vs. $(dV/dr)_b$. By employing different bob or cup sizes, the different ranges of shear rates can be studied. In the extreme, for very small annular gaps (i.e., $R_c/R_b \rightarrow 1.0$), the shear rate approaches a constant value across the annulus and can be approximated simply by

TABLE 1
Values of Calderbank-Moo-Young Shear Rate Term $(C_R)^a$

n^H	R_c/R_b ratio → 1.070	1.150	1.166	1.250	1.400	1.746
0.050	2.722	4.999	5.435			
0.100	1.708	2.617	2.801	3.735	5.184	
0.200	1.287	1.622	1.689	2.031	2.593	3.615
0.300	1.162	1.342	1.377	1.554	1.843	2.382
0.400	1.102	1.213	1.234	1.340	1.511	1.826
0.500	1.068	1.139	1.153	1.220	1.327	1.522
0.600	1.045	1.091	1.100	1.144	1.212	1.335
0.700	1.029	1.058	1.064	1.091	1.134	1.209
0.800	1.017	1.034	1.037	1.053	1.077	1.119
0.900	1.007	1.015	1.016	1.023	1.034	1.052
1.000	1.000	1.000	1.000	1.000	1.000	1.000
1.250			0.971		0.941	0.910
1.500			0.952		0.903	0.853
1.750			0.939		0.877	0.814
2.000			0.929		0.857	0.785
2.250			0.921		0.842	0.763
2.500			0.915		0.830	0.746
2.750			0.910		0.820	0.732
3.000			0.906		0.812	0.720
3.250			0.902		0.806	0.711
3.500			0.899		0.800	0.702
3.750			0.897		0.795	0.695
4.000			0.894		0.791	0.689
Coefficient (A)	0.903	0.861	1.144	0.898	1.169	1.155
Exponent (B)	−0.304	−0.505	−0.320	−0.547	−0.418	−0.485
Coefficient of fit (r^*)	0.946	0.963	0.891	0.974	0.928	0.957

[a] Each set of R_c/R_b data sets have been regressed using the least squares to a power law expression. For approximate C_R-values apply above model coefficient: $C_R = A\,(n^H)_B$.

$$\frac{dV}{dr} = \frac{2\pi RN}{R_c - R_b} \tag{28}$$

Close clearance viscometers such as this are not suitable for use with suspensions in which particle sizes are comparable to the annular gap.

End effects can be corrected for by substituting an effective length ℓ_{eff} for the true bob length ℓ in Equation 26. The effective length can be determined by calibrating the viscometer using a Newtonian fluid of known viscosity μ and applying the following formula:

$$\ell_{eff} = \frac{T}{2\pi R_b^2\, \mu (dV/dr)_b} \tag{29}$$

To correct for effective slip in rotary viscometers, the correction method of Mooney (1931) can be used. The method involves obtaining measurements using three different R_c/R_b ratios. For R_1, R_2, R_3 radii, such that $R_1 < R_2 < R_3$, three radii ratios can be defined ($S_a = R_2/R_1$, $S_b = R_3/R_2$, $S_b = R_3/R_1$). The rotational speed of the cup required to obtain the same torque for each of these cases can be measured as N_{12}, N_{23}, and N_{13}. The effective slip coefficient corresponding to the stress at the surface $2\pi R_2 \ell$ is

$$\beta_2 = \frac{2\pi^2 R_2^2}{T} (N_{12} + N_{23} - N_{13}) \tag{30}$$

Repeating this procedure for different torque values provides a correlation between β_2 and the surface shear stress τ_2. In correcting for slippage, for a given torque value, τ_b can be computed from Equation 26. The corresponding expression for wall shear stress at the cup surface is

$$\tau_c = \frac{T}{2\pi R_c^2 \ell} \tag{31}$$

The effective slip velocities at the bob and cup surfaces are $V_{sb} = \beta_b \tau_b$ and $V_{s,c}$, respectively. The effective velocity of the fluid at the rotating cup surface will be less than that of the cup by the amount $V_{s,c}$. The effective velocity of the fluid at the stationary bob surface is $V_{s,b}$. The true shear rate corresponding to τ_b can be calculated by multiplying either Equation 27 or 28 by the ratio of the actual velocity difference (i.e., from cup to bob) to the velocity difference without slip [i.e., $(V_{sc} - V_{sb})/V_{meas}$].

A modification of the concentric cylinder viscometer is the Brookfield viscometer, in which the cup radius is effectively extended to infinity. Flow curves are generated from measurements of the torque needed to rotate a cylindrical rod at various speeds when immersed in the fluid of "infinite" volume. Basically, the fluid is contained in a cup whose radius is much larger than the rod. With this arrangement, the walls of the retaining cup do not influence the shearing movement of the fluid. Equation 26 defines the shearing stress at the rod surface (R_b = radius of the cylindrical rod, ℓ = length of rod immersed in the fluid). As before, an effective rod immersion length, ℓ_{eff}, can be developed to correct for end effects. The shear rate at the surface of the rotating rod for any time-dependent fluid is given by

$$\left(-\frac{dV}{dr}\right)_b = \frac{4\pi N}{n''} \tag{32}$$

where n'' is the slope of a logarithmic plot of torque vs. rotational speed N, evaluated at the particular N for which T was measured. A plot of τ_b vs.

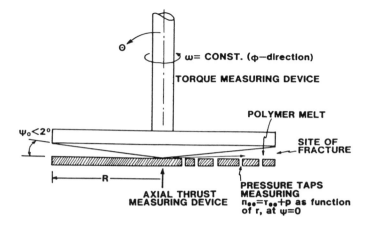

FIGURE 7. Schematic of the cone-and-plate viscometer.

$(-dV/dr)_b$ provides the flow curve. With either device a plot of $D\Delta P/4L$ vs. $8/V/D$, the Newtonian shear rate for laminar tube flow may also be generated.

In general, both devices allow studies over a very limited shear rate range. Commercial units are usually equipped with an electrical heating jacket and appropriate heat exchange to remove heat generated by friction in the fluid.

V. CONE-AND-PLATE VISCOMETER

The cone-and-plate viscometer is a widely used instrument for shear flow rheological properties studies. The device can be used to measure both viscosity and normal stresses. Applications and analysis of this type of viscometer for studying polymer melts are covered by Lobe and White (1979), Meissner (1972), and Weissenberg (1948). The principal features of this viscometer (illustrated in Figure 7) consist essentially of a flat, horizontal plate and an inverted cone, the apex of which is in near contact with the plate. The angle between the plate and cone surface is small (typically <2°) The fluid sample is placed in this small gap between the cone and plate. A flow curve is developed from measurements of the torque needed to rotate the cone at different speeds. For a constant speed N, the linear velocity at r is $2\pi rN$. The gap height at r is r tan ϕ. Hence, the magnitude of the shear rate at r is

$$\frac{2\pi\ rN}{r\ \tan\phi} = \frac{2\pi N}{\tan\phi} \tag{33}$$

The shear rate is constant between $0 > r > R$, which means that τ_{yx} is also constant over this range. For small ϕ, tan $\phi \simeq \phi$, and hence the magnitude of the shear rate equals approximately $2\pi N\phi$. The following expression defines the relationship between measured torque and shear stress:

$$T = 2\pi \, \tau_{yx} \int_0^R r^2 dr = \frac{2}{3} \pi R^3 \, \tau_{yx}$$

or

$$\tau_{yx} = \frac{3T}{2\pi R^3} \tag{34}$$

Equation 34 enables τ_{yx} to be obtained at different measured torque values. Shear rate can be obtained from Equation 33, and hence τ_{yx} can be plotted against shear rate for the flow curve.

Commercial instruments that are capable of measuring normal stresses as well as the shear stress are the Weissenberg rheogoniometer (manufactured by Sangamo Controls, Bognor Regis, U.K.) and the Mechanical Spectrometer (manufactured by Rheometrics Inc., Union, NJ).

VI. SANDWICH VISCOMETER

Another type of shear flow instrument primarily used for polymer melts is the sandwich viscometer (also known as the parallel-plate viscometer). The measurement principal is illustrated in Figure 8. A viscous material such as an elastomer is sheared between two parallel steel plates or, as in Figure 8, in a sandwich between three plates. The shear rate is

$$\dot{\gamma} = \frac{V}{H} \tag{35}$$

where H = interplate distance.

The applied force can be determined as a function of time, which after normalization with area provides shear stress data as a function of time:

$$\tau = \frac{F}{2A} \tag{36}$$

It is possible to measure transient start-up stresses of the material with this device. After a sufficiently long time and if the material is stable, the steady-state stress can be measured, at which point the viscosity can be determined as a function of shear rate. With many materials such as elastomers, both large times and equipment strains are needed. These conditions often result in slippage. To minimize slippage contact surfaces are usually knurled.

This type of viscometer is capable of making measurements at very low shear rates on materials that usually exhibit slippage in nonpressurized

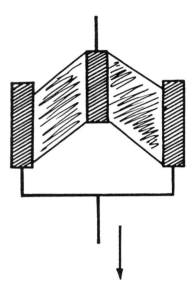

FIGURE 8. Schematic of the sandwich viscometer.

rotational devices (e.g., cone-and-plate-type viscometers). The instrument is described in detail by Zakharenko et al. (1962), Middleman (1969), Goldstein (1974), and Furuta et al. (1976).

VII. PARALLEL-DISK VISCOMETER

The parallel-disk viscometer (also known as a shearing-disk viscometer) measures viscosity in torsional flow between a stationary and a rotating disk. Design variations are illustrated in Figure 9. The operation involves the use of a serrated disk which is rotated in a sample fixed in a pressurized cavity. This system was developed for rubber applications in which both pressurization and serrated surfaces are used to avoid slippage. Properties are measured based on the shear rate at the outer radius of the disk, where

$$\gamma_a = \frac{a\omega}{H} \tag{37}$$

Shear stress and normal stress differences are given by the following relationships:

$$\tau|_a = \frac{3T}{2\pi a^3} \left[1 + \frac{1}{3} \frac{d \ln T}{d \ln \gamma a} \right] \tag{38}$$

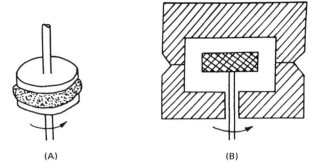

(A)   (B)

FIGURE 9.   (A) Parallel-disk viscometer and (B) Mooney shearing disk viscometer.

$$P_{11} - 2P_{22} + P_{33}\big|_a = \frac{2F'}{\pi a^2} \left[1 + \frac{1}{2} \frac{d \ln F'}{d \ln \gamma a} \right] \tag{39}$$

where F' = normal thrust. The parallel disk apparatus is limited to low shear rates.

The Mooney viscometer consists of a rotating disk in a cylindrical cavity. This instrument has become the standard quality control instrument for the rubber industry. The American Society for Testing and Materials (ASTM; Philadelphia) specifies the diameter of the disk to be 19.05 mm and a thickness of 5.54 mm. A single-point viscosity measurement is made at a standardized rotor speed of 2 RPM, after a period of 4 min and 100° or 125°C. In the elastomers industry some manufacturers still report measurements at 8 min and 127°C. Measurements are reported in terms of a calibrated torque measurement on a standard dial (referred to as ML-4 reading). The designation ML-4 means M, Mooney viscometer; L, large disk size (a smaller diameter disk, S, is used for vulcanization scorch tests); and 4, 4 min of test.

ML-4 readings can be converted to torque, shear stress, and viscosity values. Nakajima and Harrel (1979) provide the following correlations:

$$T = 8.30 \times 10^{-2} D_R \tag{40a}$$

$$\tau_a = 0.382(D_R) f' \times 10^4 \tag{40b}$$

where D_R is the Mooney unit dial reading, T and τ_a have units of N/m and Pa, respectively, f' is a function defined by the following relationship for shear stress at the disk's outer radius of the disk:

$$\tau\big|_a = \frac{T}{\pi a^3} f' \tag{41}$$

$$f' = \frac{n' + 3}{4} \left[1 + \frac{[(2/n')^{n'}(n' + 3) \, hH^{n'}]}{2a^{n'+1} \{1 - [a/(a + \delta)]^{2/n'}\}n'} \right]^{-1} \tag{42}$$

where δ = distance between the outer disk radius and cavity, h = disk thickness, and n' = power law exponent, $\simeq$ (d lnT/d ln γ_a). Function f' typically has a value of around 0.55 for the standard measurement conditions.

VIII. SLIT RHEOMETER

A slit rheometer is an instrument analogous to the capillary viscometer. The primary difference between the devices is the orifice cross-section. Han (1971, 1974, 1976) and Wales et al. (1965) describe the use of this instrument in the study of polymer melts. The device makes use of a series of flush-mounted transducers located along the flow tube. These transducers measure the pressure gradients along the direction of flow, which can then be converted to wall shear stress values via

$$\tau_w = \delta_c \frac{dP}{dz} \tag{43}$$

where δ_c = half-thickness of the channel.

Han (1976) gives the following expression for the shear rate at the channel wall:

$$\dot{\gamma}_w = \frac{3Q}{2ab^2} \left[\frac{2}{3} + \frac{1}{3} \frac{d(\ln(3Q/4ab^2))}{d(\ln \tau_w)} \right] \tag{44}$$

where a = half-width.

In general, the instrument is capable of operating over comparable shear rate ranges to the capillary viscometer.

IX. MELT FLOW INDEXER

The melt indexer is widely accepted as a standard quality control instrument in the plastics industry. The device, developed by E. I. DuPont de Nemours (Wilmington, DE) is typically used to specify polyethylenes, polypropylene, and polystyrene. It is essentially an extrusion rheometer that is in fact similar to a capillary rheometer. The instrument employs a single capillary (diameter of 0.655 mm and L/D ratio of 12) in which extrusion of material is accomplished through pressure applied by a dead weight. The melt index is given as a measure of the weight in grams extruded for a 10-min test.

The ASTM specifies a series of test conditions involving temperature and applied dead load settings. These conditions are designations A through K, and they cover the following ranges: for temperature — 125° to 275°C, for applied dead load — 325 to 21,600 g (or in terms of pressures: 0.5 to 29.5 atm). Specifications are selected so that melt indices lie between the values of 0.15 and 25 g/10 min test. The most frequently used condition in the industry is specification E(I 90°C, 2.95 atm).

Although this is a widely used instrument, its value is that of a quality control tester which can only empirically relate to end-use processability of products. The melt index (MI) is related to the inverse of viscosity; however, no end-loss corrections have been developed nor can the MI be easily related to the Weissenberg-Rabinowitsch shear rate expression.

X. ELONGATIONAL FLOW DEVICES

Up to this point instruments based on shear flow have been discussed; however, there are numerous polymer processing operations that involve uniaxial elongational flows. An example in processing in which elongational flow occurs is in the entrance die region of an extruder. This phenomenon can be studied in a capillary die. Pressure losses through such dies can be used to determine elongational viscosity in polymer melts [see Cogswell (1968, 1978), for example]. For this reason a variety of laboratory test devices and techniques have been devised over the years in order to conduct rheological investigations and/or quality control testing in this type of flow. In general, measurements of elongational viscosity are complex to interpret and more difficult than measurements made in shear flow. To make elongational viscosity measurements requires maintaining a constant elongation rate or stress while reaching steady-state conditions. These tests are typically applied to polymer melts that include high molecular weight polyolefins, styrenics, elastomers, and their compounds. A few of the most frequently used tests are briefly described below.

The most common method of measuring elongational viscosity is to stretch a filament of the material. The fluid obviously must have a sufficiently high viscosity to be maintained under controllable deformation. Materials suitable for this type of testing are elastomers and polyolefins where melt viscosities are typically 10,000 Pa/s or greater. These types of test measurements are limited to deformation rates below 5 s^{-1}. Ballman (1965), Vinogradov et al. (1970), and Stevenson (1972) describe different types of filament stretching methods. A very simple arrangement involves clamping both ends of a vertical thermostated filament, which is then stretched at a rate of dl/dt in a manner so as to maintain a constant rate of deformation. In this manner the rate of deformation ϕ is

$$\phi' = \frac{1}{\ell}\frac{dl}{dt} \tag{45}$$

and

$$\ell = \ell(o)e^{Et} \tag{46}$$

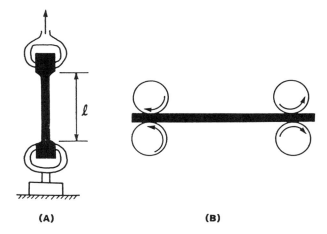

(A)   **(B)**

FIGURE 10. (A) Vertical method of elongational viscosity measurement at constant elongation rate; (B) a horizontal filament stretching test.

In other words, the length of a material undergoing stretching follows an exponential law. The test is illustrated in Figure 10A.

Another approach, illustrated in Figure 10B, involves supporting a horizontal filament on the surface of a hot, immiscible oil. The polymer filament is held at both ends between pairs of toothed or serrated wheels that turn at linear velocity V/2. The filament is drawn at both ends through the wheels. One of these wheels contains a tension-sensing instrument so that the deformation rate is proportional to the velocity ($\phi' = (V/2)/(L/2) = V/L$). Different variations of the filament stretching technique for the measurement of elongational viscosity are described by Mocosko and Lornsten (1973), Ide and White (1978), Cotten and Thiele (1979), and Dealy (1978).

XI. STRESS RELAXATION AND SMALL STRAIN MEASUREMENTS

These classes of rheological measurements are carried out on bulk polymers and are usually aimed at obtaining information on the response of the material to small strains. Strain can be applied as a step function and the stress relaxation that takes place after its application is measured.

Stress relaxation measurements are routinely made on elastomers. The test involves application of a strain to a polymer sample that is in extension, shear, or compression. The measurable parameter is the force response over time, F(t), to the applied strain. Some investigators report the material property in terms of a relaxation modulus defined by

$$E(t) = \lim_{\Delta\ell \to 0} \left[\frac{F(t)/A_0}{\Delta\ell/\ell_0} \right] \qquad (47)$$

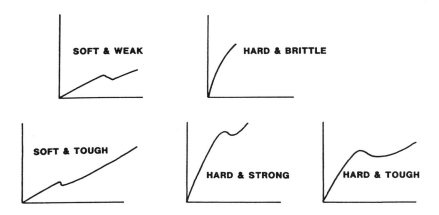

FIGURE 11.   Typical tensile stress-strain curves for polymers.

where A_0 = initial cross-section of sample; ℓ_0 = initial length of sample; and $\Delta\ell$ = amount the sample is stretched in extension.

For measurements made under shear, a surface of the sample is subjected to a displacement in a direction normal to its thickness. The shear modulus is given by:

$$G(t) = \lim_{\Delta\ell \to 0} \left[\frac{F(t)/A}{\Delta\ell/H} \right] \tag{48}$$

where H = sample thickness.

These types of tests are normally coupled with mechanical properties testing. A variety of polymers, particularly elastomers, are widely used in consumer- and industry-related products because of their good mechanical properties. Tensile measurements at several strain rates are often made as a quality control check of the integrity of the polymer, even though their behavior in compound states may be quite different. The stress-strain curve provides information on both the rheological and mechanical behaviors of a polymer. Figure 11 illustrates typical features of tensile stress-strain curves for polymers. The classification of curves in Figure 11 shows that polymers can be classified in general terms as rubbery, glassy, or variations in between. Compounding can greatly change the stress-strain response curve of a given polymer. Tensile testing in rubber laboratories is normally done on a tensile testing machine (such as an Instron) and performed according to ASTM standard methods (ASTM D638, ASTM D882). As a product quality control test, many elastomer manufacturers are no longer using tensile testing of neat rubber, but rely more heavily on rheometer-cure-response data. In product development work, however, tensile testing is very much an integral part of polymer characterization work. A minimal testing scheme involves obtaining

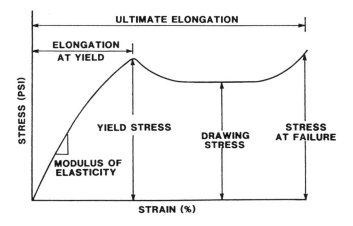

FIGURE 12. Tensile stress-strain curve for plastics.

a stress-strain curve at room temperature conditions and at perhaps three or more strain rates. From each curve, various quantities related to the mechanical and rheological properties of the sample can be derived. Figure 12 shows the generalized tensile stress-strain curve for plastics and the various parameters of interest. The principal information obtained from these curves includes:

1. The modulus of elasticity (the initial slope of the curve)
2. The yield stress (peak stress in the early portion of the curve)
3. The stress at failure
4. The elongation (or strain) at yield (also called the yield strength)
5. The ultimate elongation or strain at failure (also referred to as the elongation at break)
6. The energy to break, measured by the area under the stress-strain curve up to the point of failure

Feature 6 is truly a measure of sample toughness.

An important class of experiments originally introduced by Philippoff (1934) involves the determination of stress response to an imposed sinusoidal oscillation. Results are analyzed in terms of the dynamic viscosity function (not to be confused with the dynamic viscosity of classical fluid mechanics) and the dynamic rigidity function G'. In oscillatory shear flow, we may represent the velocity field using rectangular Cartesian coordinates as follows:

$$V_1 = \epsilon \, \omega \, \chi_2 \, e^{i\omega t} \quad , \quad V_2 = V_3 = 0 \tag{49}$$

where ϵ is small enough for second- and higher-order terms to be neglected and is frequency. The corresponding stress field is

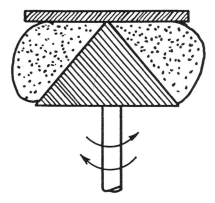

FIGURE 13.   Oscillatory shear in a cone-and-plate device.

$$P_{12} = \eta^*(\omega) \, \epsilon \, \omega \, e^{i\omega t} \quad ; \quad P_{11} - P_{12} = P_{22} - P_{33} = 0 \qquad (50)$$

where η^* is the complex dynamic viscosity expressed by convention as:

$$\eta^* = \eta' - i\,\frac{G'}{\omega} \qquad (51)$$

Other functions typically defined are

$$\eta'' = \frac{G'}{\omega} \qquad (52a)$$

$$G'' = \omega \eta' \qquad (52b)$$

The most widely used method for measuring $G'(\omega)$ and $G''(\omega)$ on polymer melt systems involves placing the melt in the gap between a cone and plate and oscillating one of the elements relative to the other with a shear strain. Figure 13 illustrates this technique.

Another arrangement, illustrated in Figure 14, involves rotating two parallel eccentric disks. The polymer sample undergoes periodic sinusoidal deformation. The forces exerted on the disk are interpreted as $G'(\omega)$ and $G''(\omega)$.

Many industrial research laboratories use the Weissenberg-Rheogoniometer and Rheometrics Mechanical Spectrometer for cone-and-plate sinusoidal oscillation measurements of $G'(\omega)$ and $G''(\omega)$. The reader is referred to the works of Vinogradov et al. (1972), Lobe and White (1979), Munstedt (1976), Aoki (1979), Andrews et al. (1948), Tobolsky (1960), Maxwell and Chartoff (1965), Chen and Bogue (1972), and Walters (1975) for discussions of sinusoidal experiments.

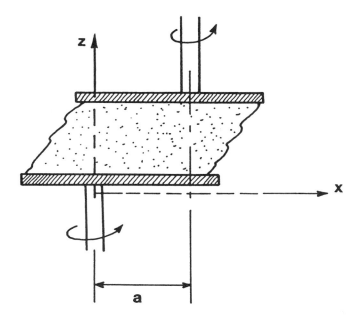

FIGURE 14.   The principle behind the orthogonal rheometer.

XII. FLOW BIREFRINGENCE MEASUREMENTS

Determination of birefringence during flow is a rheological property measurement technique applied almost exclusively to homogeneous systems of flexible polymer chains. This technique is based on a relationship between flow birefringent behavior and stresses and is determined by the rheo-optical law established by Philippoff (1956) and others (Dexter et al., 1961). The law relates the difference in the principal refractive indices to the difference in the principal stresses as developed during flow. The principal stresses represent the magnitudes of the principal axes of the ellipsoid of the stress tensor. The technique has been applied in purely scientific studies on flexible chain polymer melts such as polyolefin, polystyrene, and elastomers. The major difficulty in its application is that the technique is limited to amorphous polymers. Even trace amounts of crystallinity have a dramatic effect on measurements. Measurements can be performed on polymer solutions using a coaxial cylinder viscometer, whereas polymer melt measurements are normally taken on parallel-plate slit extrusion devices. These instruments seek to use flow birefringence to measure shear flow properties.

Measurement of birefringence is accomplished by studying the transmission viscometric of a polarized monochromatic light beam that is passed through the fluid medium to some depth. Since the medium in the plane perpendicular to the transmitted wave has different refractive indices in different directions, the wave splits into two parts. Each refracted wave has a different velocity that corresponds to the two refractive indices. The waves are out of phase with each other and the beam exiting the instrument passes through a polarizer crossed normal with the initial direction of polarization. Birefringence is determined either through the reduction in intensity of the exiting beam, through the use of compensators, or through characterization of the fringe patterns produced by the external polarizer. The fringes indicate conditions under which the contributions of the two waves to the direction allowed passage by the analyzer cancel. The method based on the use of compensators is described by Wales and Philippoff (1973). Classical theory and application of flow birefringence measurements of rheological properties are treated in the following references: Wales (1969), Han and Drexel (1973), Brizitsky et al. (1978), Andrews et al. (1948), Wales and Janeschitz-Kriegle (1967), Gortemaker et al. (1976), Meissner (1969), Dexter et al. (1961), Lodge (1956), Chen and Bogue (1972), Gortemaker and Janeschitz-Kriegle (1976), and Treloar (1940, 1941, 1947).

XIII. SWELL AND SHRINKAGE TESTS

Rheological test more often used for product quality control is a measurement of the elastic recovery in polymer melts and their compounds. This is usually accomplished by measuring the swell (or shrinkage) of materials undergoing extrusion. The tendency of polymers, whether thermoplastic or rubber, to enlarge when emerging from an extruder die is called die swell. Die swell normally refers to the ratio of extruded size to die size. This behavior is a measure of the relative elasticity in the flowing polymer stream. Die swell is caused by the release of the residual stresses when the sample emerges from the die. Measurement of this behavior has become widely recognized in the rubber and plastics industries as an important indication of polymer processability.

Traditional measurements reported by various investigators have been made on extrudates either with a micrometer or by weight per unit length. The Monsanto Industrial Chemicals Co. (Akron, OH) has introduced an automated capillary rheometer (the Monsanto Processability Tester) which employs an extrudate swell detector based on interaction with a scanning laser beam positioned immediately below the capillary die exit. This enables measurement of the running die swell. Figure 15 shows running die swell data for a compound correlated against wall shear for different L/D dies. It is important to note that with some materials a distinction between running die swells and relaxed die swells should be made. Highly elastic materials will tend to swell more over time depending on their elastic recovery energy.

C. W. Brabender Instruments, Inc. (Hackensack, NJ) markets an optical, electronic die swell measuring system that can be used for continuous control and monitoring of extruders and plasticorders. The basic design features are

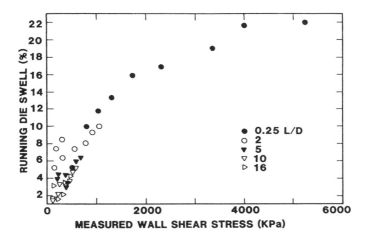

FIGURE 15. MPT data on running die swell.

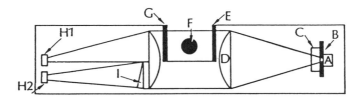

FIGURE 16. Schematic of optical-electrical die swell tester.

illustrated in Figure 16. An infrared, long-life diode (A) emits homogeneous infrared light through a filter (B), diffuser (C), and lens (D); then through a window (E). The light impinging on the object (F) is scattered. Only that light reaching the ambient light filter (G) is measured by the photodetector H1. Light reaching photodetector H2 is diverted through the prism (1) and is uninterrupted. This acts as a reference beam. The electronic instrumentation measures the differences between photodetector H1 and H2 and calculates the object thickness to an accuracy of 0.2% of full scale. The object to be measured must be located vertically within the measuring area. The maximum permissible deviation from vertical is 2.5°. Operation at angles >2.5° from vertical will yield greater errors than 0.1%. If the measuring angle is >2.5° from vertical, the measurement will be increased by the factor 1/cos a, where a is deviation from vertical.

While the upper collimated light serves as the measuring beam, the lower portion impacts on the second photodetector and acts as the reference beam. Thus, the temperature and age affected by drift which occurs on all semi-conductor elements is negated.

In some applications such as hose and tubing manufacturing, it is important to know both the extrudate diameter and cross-section. The device described above can provide two diameter measurements, which will give an out-of-round indication (eccentricity).

Experimental studies reported in the literature using various methods of swell measurement are reported by Lobe and White (1979), Graessley et al. (1970), Pliskin (1973), and White and Roman (1976).

XIV. VISCOMETRIC TECHNIQUES FOR DIFFERENT INDUSTRIES

A variety of other instruments are employed by various industries for either product development work or product quality control. Although these methods shall be reviewed in relationship to the industries and products they are most frequently applied to, these devices are by no means restricted to those applications. One type of instrument may have become a standard tool for a particular industry largely through an evolution of preference and in some cases convenience, rather than primarily for technical reasons. Most of these techniques can be applied to any fluid characterization; however, the techniques can greatly vary in terms of the specific information they provide.

One industry in which rheological characterization has been applied for many years in a subjective manner is the manufacture of liquid detergents. Consumer applications for these products include dishwashing, hard surface cleaning liquid scourers for walls and floors, fabric softeners, thickened bleaches, liquid abrasive cleaners, combination liquid detergent-wax products for cars, and liquid detergents for washing fabrics. Other examples include personal washing products such as hair shampoos and conditioners, body shampoos, shower gels, toothpaste, and skin cleansers. The particular rheological properties imparted to these various products are governed in a number of cases by consumer preference or attitudes as well as by technical benefits derived from certain flow properties. The consumer often makes his or her own rheological assessment through sensory perception (touch and sight) of the product. The consumer describes different products as watery, thin, thick, jelly-like, paste-like, etc. It is the rheologist's job in part to be able to quantify this subjective terminology so that product development can meet the consumer's desires.

Among the most frequently used viscometers in this industry are the U-tube viscometer, the rolling ball (or Hoeppler viscometer), the Ford cup, and the Brookfield viscometer, illustrated in Figure 17. The first three devices have often been used in a totally empirical manner for quality control purposes. For example, in each of these cases the time for the material to flow through the tube or from a vessel, or for a ball to roll between two points, would be measured over some conveniently specified time interval. With some devices a scale might be included to display a reading of variables, as in the case of

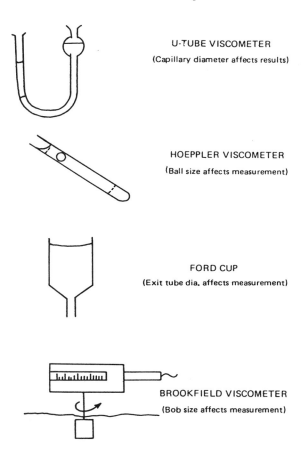

FIGURE 17.　Various viscometers employed in the detergent industry.

the Brookfield, showing readings between (for example) 10 and 90% of full-scale deflection. Many devices have been adopted, not only in this industry, on the basis of convenience and expediency rather than on a scientific or an engineering basis. A disadvantage of many of the simpler instruments is their inability to generate more than one value to characterize a fluid. Often this reading is converted into a viscosity value based on a Newtonian calibration.

Barnes (1980) describes several modifications to the various techniques used in characterizing detergents. The first of these modifications concerns the Hoeppler viscometer, which normally operates using a fixed tube angle and several balls. He proposes a variable angle of rolling. By altering the angles, the average stress in the fluid can be made to vary because of the change in magnitude of the gravitational force on the ball in the tube direction. For this arrangement, viscosity data can be fitted to a relation:

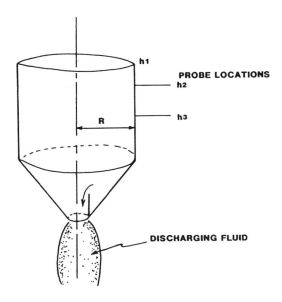

FIGURE 18.   The operation of the Ford cup.

$$\eta = k_1 (\sin\alpha') \theta(A + Bn) \qquad (53)$$

where α' = angle of the tube from the horizontal; θ = time for the ball to roll between two fixed points; and n = slope of curve generated by plotting In $(\sin \alpha')$ vs. $n(l/\eta)$.

Note that shear rate in this experiment is $\dot{\gamma} = k_2/\theta$, and A, B, k_1 and k_2 are constants. Also, A + B = 1. Usually two experiments (i.e., tests at two angles) provide sufficient information such that a power law relation can be derived from η values.

The Ford cup provides a measure of the time for a fixed volume of fluid (e.g., 50 ml) to flow from the cup into a graduated cylinder. One problem in using this technique with such materials as detergents is the formation of foam, which can make accurate flow time determinations difficult. This can be overcome by employing electroresistivity probes along the inside walls of the cup (see Figure 18). This provides accurate measurement of the flow times involved. A sharp edge orifice fitted to the outlet tube provides steady calibrated flow conditions. Barnes (1980) gives the following expressions for viscosity and average shear rate using this technique:

$$\eta = \mu\left(\frac{1 + n}{2}\right) \quad , \quad N - s/m^2 \qquad (54)$$

$$\gamma_{ij} = \frac{2(h_i - h_j)}{\theta_{ij}} \frac{R^2}{r^2} \quad , \quad sec^{-1} \tag{55}$$

where i, j refer to different heights h_1, h_2, h_3, . . . , etc.

Measurements using a modified Ford cup are quite comparable to those obtained with concentric cylinder viscometers. Materials that are highly thixotropic generate flow times which can be much longer than predicted if steady-state concentric cylinder data are used (Barnes, 1980) for comparison, because the total shear in the Ford cup is small compared to the steady-state rotational data. The Ford cup is not applicable to materials in which elongational viscosity is much greater than shear flow at the same shear rates. Many detergents show anomalous elastic effects and hence the Ford cup is not suitable for viscosity measurement in all cases.

Viscosity measurements in shear flow of detergent products can be made in commercial rheometers based on cone-and-plate instruments or concentric cylinder viscometers. Typical commercial units employed are the Haake Rotovisko, the Ferranti-Shirley viscometers, and the Contraves Rotary viscometer. These instruments cover shear rates typically between 1 to over 1000 s^{-1}. Other instruments employed to a lesser extent are the Deer rheometer and the Weissenberg Rheogoniometer.

The Deer rheometer is designed to subject the sample to a constant stress whereby the subsequent deformation is measured. This instrument uses a frictionless air bearing so that it is capable of applying a variety of stress patterns with variable time of application and rate of application of stress. Variations of this instrument are capable of applying oscillatory stress to the fluid sample. Deformation can be studied to measure several parameters: creep, steady-state shear rate, recoverable shear, and various oscillatory parameters. Through geometry changes the instrument can cover a range of applied stresses. Barnes (1980) describes a modified Deer rheometer employed in the study of soap rheology. The system is illustrated in Figure 19 in which the rheometer has been modified to measure the extensional properties by simple bending performed at very small deformation. Torque is applied to two vertically mounted rods of soap that are fixed at both ends. Torque is applied to the samples through the sample end retaining sleeves by magnetic knife edges mounted on radial arms. The knife edges ensure that no twisting of the rods takes place during bending. The extensional compliance is calculated from bending beam theory analysis.

The Weissenberg Rheogoniometer, as noted earlier, can be used to measure G' and G'' as a function of frequency and deformation. These parameters can also be measured for a range of superimposed values of steady shear rate. It is most often applied in industrial laboratories at measurements over a range of shear rates and as a function of shear/time at initiation or cessation of shear. Other frequently used viscometers in the detergent manufacturing industry include capillary and orifice viscometers and penetrometers. As noted

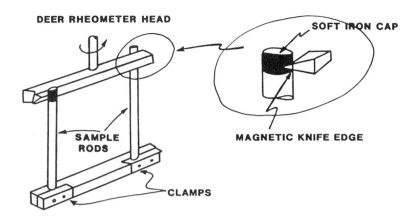

FIGURE 19. A Deer rheometer.

earlier in this chapter, a capillary viscometer operates by forcing the fluid through a straight circular tube. This instrument is capable of shear rates up to 10^5 s^{-1}. For fluids obeying power law behavior, the Mooney-Rabinowitz correction makes possible a definition of the average viscosity-average shear rate parameters. Orifice viscometers are often used in the plastics industry to measure parameters important in pumping fluids through circular orifices. The detergents industry has applied such devices to handle very viscous semisolid materials such as soap.

The last noteworthy device is the penetrometer, which monitors the movement of a weighted penetrator into the surface of a very viscous sample. The penetrometer is most often used for testing such materials as butter or bitumen, in which the penetrator is a sharp, truncated pin. Within the detergents industry, this device has proven useful for semisolid soaps.

A variety of other instruments and/or variations of devices described above are available that are applied in lubrication technology. Lubricating oils are manufactured from the distillation of crude oil. Primary distillation is performed at atmospheric pressure to remove the lighter gasoline, kerosene, and gas oil fractions, and the lubricating oils remain in the residue that boils above about 370°C. The residue must be distilled under high-vacuum conditions, resulting in several lubricating-oil fractions and an asphaltic bitumen. The second-stage residue from those crudes low in asphalt is also a source of lubricating oils. These fractions are referred to as residual oils to distinguish them from distillate oils, which are boiled off in a high-vacuum still. Lubricating-oil fractions can constitute up to 5% of the total crude, depending on the crude source. A further refinement step in manufacturing the lubricating oils is usually performed with acid or solvents to remove a variety of unwanted constituents such as sulfur compounds. Hydrotreating at moderate temperatures and pressures saturates the unsaturated hydrocarbons and removes sulfur-bound molecules. Paraffinic constituents separate out as wax at low temperatures. These are removed by chilling and filtering or by a solvent treatment process. Polar compounds and dark coloring constituents are removed from the oil in a final treatment stage using active clays. The refinery streams in these processes are referred to as base oils in the petroleum and allied industries. Base oils are characterized by the origin of the crude, the specific refining process used, viscosity, their viscosity index (VI), and various other properties. The VI is a measure of the variation of viscosity with temperature, and, in general, the higher the VI the smaller the variation.

An important class of products is mineral oils, examples of which are paraffins and aromatics. Oil additives are usually supplied as concentrates in mineral oil and are readily added during blending operations. Additives can have a variety of purposes for the end-use product, for example, oxidation inhibitors, additives to neutralize acids, surface-active agents to disperse and maintain other additives in suspension, ultrahigh-pressure lubricating additives. Polymers are also added to oils to modify rheological properties at extreme temperature conditions (e.g., for motor oils, to increase pouring ability at very low temperatures). A general example of a pour-point depressant is an additive that inhibits the formation of a wax-crystal network at lowtemperatures. This type of product extends the liquid property range of the lubricant. In lubrication technology, the Navier-Stokes equations for incompressible, viscous inelastic fluids are often applied to engineering problems. Information on viscosity and proper application of these equations is usually sufficient to estimate oil film thickness and the friction. These equations are usually solved isothermally for computation simplicity and do not account for viscous heating effects. Hence, it is necessary to fully define the viscosity-temperature dependence of these products in order to correct engineering formulations from the Navier-Stokes equations. This is done, as noted earlier, through the use of the VI; standardized procedures are described in "Kinematic Viscosity", ISO Method 3104, IP Method 71, ASTM Method D445, and "Industrial Liquid Lubricants", ISO Viscosity Classification - ISO 3448.

The capillary viscometer is the standard instrument used for these measurements. Temperature conditions are standardized in these test methods so that oils can be compared on a relative basis (e.g., 40° and 100°C). In the ISO Viscosity Classification system, a 10-grade oil has a viscosity between 9.0 and 11.0 cSt at 40°C. In total there are six grades for each decade in viscosity.

Many commercial oils contain polymeric additives. Adding small amounts of polymer to an oil reduces the rate of decrease in viscosity with increasing temperature. Examples of the types of polymers used are polyisobutene, polyalkylmethacrylates, and copolymers such as styrene-butadiene and styrene-isoprene. Molecular weights of these additives range from about 50,000 to several hundred thousand. In an automobile engine, motor oil is exposed

TABLE 2
SAE Viscosity Grades for Engine Oils

SAE viscosity grade	cP at $-18°C$ (ASTM D 2602) (Max.)	Viscosity range	
		cSt at 100°C (ASTM D 445)	
		(Min.)	(Max.)
5W	1,250	3.8	—
10W	2,500	4.7	—
20W	10,000	5.6	—
20	—	5.6	<9.3
30	—	9.3	<12.5
40	—	12.5	<16.3
50	—	16.3	<21.9

to effective shear rates ranging from 10^4 to 10 s^{-1} over wide temperature ranges. Considerably lower shear rates are involved during engine starting. In starting an engine, the cranking speed must be high enough to enable a cylinder to fire. Furthermore, the power the engine delivers when it does fire and the assistance of the starter motor are disengaged during this process. Both these actions depend on the viscosity of the motor oil. One important industry test which is a low-temperature, high-shear test method is the Cold-Cranking Simulator (ASTM Method D2602). The U.S. Society of Automotive Engineers (SAE) provides a classification of engine oil viscosity based on this test method. Table 2 lists this classification, which is based on a low-shear viscosity measurement at 100°C and a high-shear viscosity measurement at $-18°C$. It should be noted that the industry is in the process of revising test methods and the classification system. Current evolution is toward a high-shear cold-cranking simulator (CCS) viscosity and a yield stress value to be obtained in terms of specified temperatures.

The design of motor oil products requires detailed knowledge of the rheological demands on the oil made by the engine. For example, high temperatures normally achieved in an automobile engine (e.g., 100° to 150°C) causes shear thinning, which can result in increased wear. Also, however, at low temperatures shear thinning provides the advantage of easier starting conditions. It was noted earlier, however, that base oils are dewaxed in order to remove paraffinics which crystallize out of solution at low temperatures. Engine oils, however, are never completely dewaxed because of prohibitive refining costs. Consequently, commercial oils can encounter temperatures in which residual waxes will separate out. This important property is tested for by the so-called cloud point, based on ISO Method 3015, IP Method 219, ASTM Method D2500. The cloud point is the temperature at which wax first begins to separate. Another rheological test of motor oil products is the pour point (ISO Method 3016, IP Method 15, ASTM Method D97). At lower temperatures oils acquire a yield stress due to the network structure formed

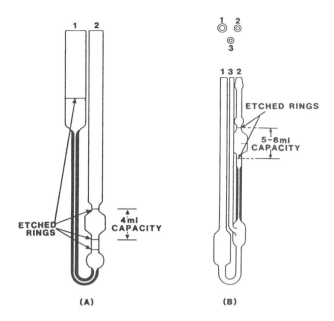

FIGURE 20. (A) Reverse-flow capillary viscometer; (B) suspended level viscometer.

by the wax crystals. Considerable research effort has gone into the development of pour-point depressants, which are additives that interfere with the formation of wax crystal structure.

Illustrations of viscometric instruments used in this industry are given in Figures 20 through 24. Figure 20 shows a reverse-flow capillary viscometer. This instrument is used for dark, opaque oils. For this device the oil flows upward past a set of "timing marks". In this manner the position of the fluid miniscus is not obscured by oil left behind on the glass wall. Figure 20B illustrates a suspended-level viscometer.

Viscosity measurements at low temperatures for transmission and gear oils are usually made with a Brookfield viscometer (methods IP/267 and ASTM D.2983). Other devices equally applicable are the Haake Rotovisko RV2 (i.e., constant-speed concentric-cylinder instrument), the Hoeppler rolling ball viscometer, and the U-tube capillary viscometer.

The cold-cranking viscometer cell is illustrated in Figure 21. This device consists of a concentric-cylinder viscometer. A constant-temperature bath is controlled to − 18°C and the oil sample is injected into the viscometer volume via a syringe.

Several other viscometers specifically developed for lubrication research are illustrated in Figures 22 and 24. Brief descriptions of these instruments are provided in Table 3 along with techniques applied to rheological characterization of various products from other industries.

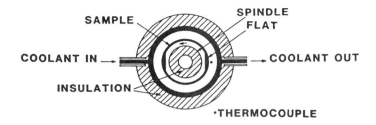

FIGURE 21.   Schematic of cold cranking simulator viscometric cell.

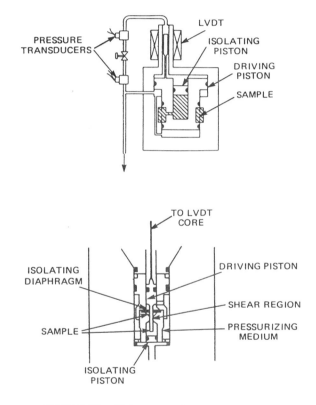

FIGURE 22.   High-pressure shear stress apparatus.

A number of other techniques have evolved in different industries that serve as both scientific devices for product characterization and as single-point measurement quality control techniques. Principal instruments are summarized in Table 3 according to industry preference. Key references describing the theory and application for that particular industry are included.

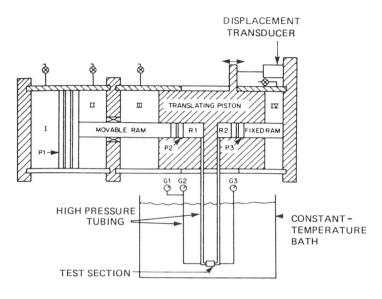

FIGURE 23. A high-pressure capillary viscometer described by Novak and Winer (1963).

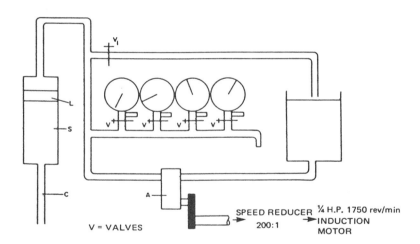

FIGURE 24. A grease viscometer based on ASTM methods.

TABLE 3

Summary of Major Viscometric Techniques Used Throughout Different Industries

Products	Type of viscometer	Operating principles — rangeability	Standard test methods	Suggested ref.
Lubricating oils	Glass capillary viscometers	Viscosity — shear rate measured in capillary tubes of different L/D, over temperature range. Figure 27 shows modified capillary where test fluid contained in reservoirs R1 and R2 and interconnecting tubing. Material separated from hydraulic fluid contained in I–IV. Adjusting press, between III and IV, translating piston drives test fluid through capillary. Shear stress range is 30 Pa-5 MPa.	ISO Method 3104; IP Method 71; ASTM Method D445; ISO Viscosity Classification ISO 3448	Murphy, C. M. et al. *Trans. ASME*, 71, 561 (1949).
	Cold cranking simulator	Concentric cylinder viscometer with temperature control bath; measurement made at constant power, not constant shear stress or shear rate.	ASTM Method D2602	
	Bridgman falling-body rheometer	Measures viscosity at low shear rates and high pressures. Time measured for plummet of known weight to fall known distance vertically through oil. Manual timing limits method to viscosity ranges of 10^{-3}–10^3 Pa/s; accuracy $> 5\%$.		Irving, J. B. and A. J. Barlow, *J. Phys. E. Sci. Instr.*, 4, 232 (1971).
	High-frequency oscillatory measurements	Measures shear modulus (G) and time delay from instant of application of high contact press, to attainment of equilibrium specific volume. Technique based on measurement of change in amplitude and phase of a pulsed train of mechanical waves on reflection at the interface between fuze quartz and oil. Waves are generated piezoelectrically.		Hutton, J. F., M. C. Phillips, J. Ellis, G. Powell, and E. W. Jones, in *Electrohydrodynamics and Related Topics*, C. M. Taylor, M. Gudet, and D. Berthe, eds., Mechanical Engr. Publ. Ltd., Bury St. Edmunds, 1979.

	Optical techniques	Measurement depends on existence of Debye elastic waves caused by thermal motions of molecules. The waves display acoustic properties, containing Debye waves which are propagating, hence the Doppler shift of frequency of light probe source is detected. Scattered light can be detected by an interferometer. Doppler frequency shift gives velocity of Debye waves. Oil sample must be near glassy state for this technique.	Harrison, G., *The Dynamic Properties of Supercooled Liquids*, Academic Press, London, 1976. Matheson, *Molecular Acoustics*, Wiley, London, 1971. Alsaad, M. A., W. D. Winer, F. D. Medina, and D. C. O'Shea, *Trans. ASME (F) J. Lubr. Tech.*, 100, 418 (1978). Bair, S. and W. D. Winer, *Trans. ASME (F) J. Lubr. Tech.* 101, 251 (1979) Dill, J. F., P. W. Drake, and T. A. Litovitz, *ASLE Trans.*, 18, 202 (1975).
Lubricating greases	Weight penetration method	Measure of grease yield stress in terms of consistency. Basic test for consistency by dropping a specified cone from a given height into sample. After sinking time of 5 s the cone is stopped and depth of penetration measured in units of 10^{-4} m. Greases classified according to penetration determined at 25°C.	Penetration of greases — IP Methods 50 and 310; ASTM Methods D217 and D1403
	Capillary viscometer, also Shell-de Limon viscometer	Measure of apparent viscosity. Data generate standard rheological flow curve (i.e., wall shear stress vs. shear rate). At low shear rates existence of yield stress observed. At high shear rates curve approaches line of zero slope, indicative of the viscosity of the base oil from which grease is derived. Basic system described by ASTM Method D1092. Lowest shear rate range is 10 s^{-1}.	"Apparent Viscosity" — ASTM Method D1092. "Flow Properties at High Temperatures" — ASTM Method D3232. Klamann, D., L. Endom, R. Rost, and A. Haak, *Erdol Kohle*, 20, 219 (1967).

TABLE 3 (continued)
Summary of Major Viscometric Techniques Used Throughout Different Industries

Products	Type of viscometer	Operating principles — rangeability	Standard test methods	Suggested ref.
	Cone-and-plate viscometers	Operation described earlier in this chapter. To avoid slippage, plates are usually grooved or roughened. Close spacing used between plates so that main portion of shearing boundary is seen by grease between the grooves.		Hutton, J. F. in *The Rheology of Lubricants*, T. C Davenport, Ed., Applied Science Publishers, Barking, 1973.
	Torsional balance rheometer	Provides normal stress measurements. Operates by applying a known normal force to parallel plates between which sample is held. Plates are of fixed radii, of which one is free to move along axis joining centers. The rotation is applied and the equilibrium separation of plates denotes the equilibrium of the applied force with normal force generated by the grease sample at the measured shear rate.		
Organic coatings Paints Printing inks	Consistency cups	This is a single-point measurement instrument. Flow ability measured by gravity-induced flow through an orifice. Paint standards require use of a conical bottom cup (Ford cup or ISO cups). Some designs employ a spherical bottom (Zahn and Shell cups). Efflux times provide measure of viscosity, which in part depends on the orifice geometry. High aspect ratios (orifice *L/D*) improve smoothness of flow and the linearity of the time-viscosity correlation. Viscosity measurement range can be extended by changing the orifice dimensions.	ASTM D1200-70; ISO 2431-1972	Westgate, M. W., *Off. Dig. Fed. Soc. Paint Technol.*, 32, 616 (1960). McKelvie, A. N., *Progr. Org. Coatings*, 6, 49 (1978). Walters, K. *Rheometry*, Chapman & Hall, London, 1975, p. 56. Patton, T. C., *Paint Flow and Pigment Dispersion*, John Wiley & Sons, New York, 1964.

Falling-sphere rheometer	The technique is best suited for materials that approach Newtonian behavior. The net velocity of a solid sphere submitted to buoyancy and viscous friction forces is measured. From Stokes law dynamic viscosity can be computed from the measured time of fall t of the sphere: $(g/4.5)(s^{-1})r_2(l/t)$; where g = gravitational acceleration, l = distance sphere falls, r = sphere radius. Equation ignores wall effects from the container. Corrections for wall effects can be made by a factor which accounts for tube-to-container radius: $(g/4.5)(s^{-1})r_2(l/t)[l - 2.104(r/R) + 2.09(r/R)_3 \ldots]$. By using spheres of different radii and/or densities, the range of viscosity measurements can be extended. The technique is not applicable for thixotropic fluids (some paints are such fluids).	
Rising-bubble rheometer	The method is essentially the same as the falling-sphere method. Buoyancy works in the same direction as for spheres, but with lower density than that of the test fluid. Hence, principle is based on time for bubble to rise through a column of fluid. Since the bubble is not truly spherical, the above equations do not apply rigorously. Relating bubble rise time to viscosity must therefore be based on calibration of the technique using fluids of known viscosity.	ASTM D1545-63

Patton, T. C., *Paint Flow and Pigment Dispersion*, John Wiley & Sons, New York, 1964.
Gardner, H. A. and G. G. Sward, *Physical and Chemical Examination of Paints*, Gardner Laboratories, Bethesda, MD, 1962. |

TABLE 3 (continued)

Summary of Major Viscometric Techniques Used Throughout Different Industries

Products	Type of viscometer	Operating principles — rangeability	Standard test methods	Suggested ref.
	Rolling-sphere test and sliding-plate test	Usually applied to measure viscosity changes in drying paint films. *In situ* measurements are made by either sliding, rotating, or rolling objects over the film surface and tracing the speed as a function of time. Film/substrate are usually inclined to some defined angle. These methods operate at given stress levels as opposed to a given shear rate. Generally, these provide a basis for a quality control test and/or for relative comparisons between samples.		Fischer, E. K., *J. Colloid Sci.*, 5, 271 (1950). Goring, W. and N. Dingerdissen, *Farbe Lack*, 83, 270 (1977). Kornum, L. O., *Fatipec XIV*, V. Takacs, Ed. Budapest, 329 (1978). Kornum, L. O., *Rheol. Acta*, 18, 178 (1979). Quach, A. and C. M. Hansen, *J. Paint Technol.*, 46, 40 (1974).
	Couette-type viscometers	Couette-type (coaxial cylinder geometry) viscometers are widely used for characterizing pigment dispersions. Since the sample for rheological testing is a dispersion, it is important that the tolerance between the cylinder be much larger than particle and aggregate dimensions. As a rule of thumb, the gap should be a factor of 10 larger. Various factors that can affect measurements are sedimentation of pigments, volatility of sample, pigment migration, wall slippage, entrapment of air with very viscous samples, and viscous heating. Pigment migration often translates into highly thixotropic behavior and it is therefore advisable to repeat measurements		Middleman, S., *The Flow of High Polymers*, Interscience, New York, 1968. Mewis, J., *J. Non-Newtonian Fluid Mech.*, 6, 1 (1979). van Wazer, J. R., J. W. Lyons, K. Y. Kim, and R. E. Colwell, *Viscosity and Flow Measurement*, Interscience, New York, 1963. Karnis, A., H. L. Goldsmith, and S. G. Mason, *J. Colloid Interface Sci.*, 22, 531 (1966).

several times to ensure that thixotropy actually characterizes the product or if it is associated with the present state of the sample. One serious source of error can be introduced by using very large gaps between cylinders to accommodate dispersion; this leads to nonuniform shear over the gap which cannot be corrected for easily.

Cone-and-plate viscometers

Advantages over Couette-type viscometers include small sample size required, homogeneous shear rate, and possibility of immediate conversion of data to apparent viscosities. Similar problems of data inaccuracies as outlined for Couette-type instruments apply. Shear fracture of printing inks has been observed in cone-and-plate instruments. It is important that preliminary experiments be run to define at what shear rates this phenomenon occurs.

Klijn, P.-J., J. Ellenberger, and J. M. Fortuin, *Rheol. Acta, 18, 303* (1979).
Cheng, D. C.-H., J. Appl. Phys., 17, 253 (1966).
Hadjistamov, D. and K. Degen, *Rheol. Acta*, 18, 168 (1979).

Stormer-Krebs viscometer

Rotating-element viscometer which employs a two-bladed paddle. The device is most often used for product quality control testing and is operated at constant torque as opposed to constant speed as with other rotating-element viscometers. The constant torque is controlled by a weight W which causes the paddle to rotate at a given speed of 200 RPM, providing a measure of viscosity. The viscosity value is obtained from a weight-rotational speed diagram. Tests can be accelerated by the use of a stroboscopic instrument. Data are typically reported in arbitrary values of Krebs units

Epprecht, A. G., *Farbe Lack*, 84, 86 (1968).
Camina, M., *Defazet*, 28, 117 (1974).
Freier, H. J., *Farbe Lack*, 69, 87 (1963).

TABLE 3 (continued)
Summary of Major Viscometric Techniques Used Throughout Different Industries

Products	Type of viscometer	Operating principles — rangeability	Standard test methods	Suggested ref.
		(KU). The manufacturer provides a conversion table to W-KU. In the paint industry the device is normally applied as a single-point measurement device. It is, however, possible to generate a shear rate distribution by changing weights.		
	Falling-rod viscometer	As in the Couette geometry the test sample is held in an annular gap; however, one of the cylindrical walls (usually the outer one) moves under gravity while the other is stationary. This results in a telescopic flow field. The device has similar advantages to a cone-and-plate viscometer. The shear rate range can be extended by applying different weights to the rod. The actual measurement is the time for the rod to fall over a specified distance. Shear stresses can reach up to 10^4 N/m² with viscosities up to 100 Pa/s at shear rates as high as 10^3 s⁻¹.		Turner, T. A, *Br. Ink Maker*, 15, 59 (1973).
	Band viscometer	Based on the same principle as the falling-rod viscometer, except that a band moves through a slit filled with the test sample. Shear rate ranges can be extended by using different weights attached to the lower end of the band.		Taylor, J. H. and S. L. Cozzens, Natl. Printing Ink Res. Inst., Bull. No. 53, 1959. Patton, T. C., *Paint Flow and Pigment Dispersion*, John Wiley & Sons, New York, 1964.

Instrument	Applications	Description	References
Rotary tackmeters		Employed in the printing ink industry. Device consists of three rotating rollers, with the central roller driven by a motor. A second roller is positioned on top of the central roller, and the third roller provides reciprocating motion that smooths the liquid layer before it enters the nip between the other two rollers. The fluid sample transmits forces from the surface of one roller to the other. The external force required to maintain the roller in place provides a measure of the tack. These data can then be related to viscometric data.	Dobbels, F. and J. Mewis, *Chem. Eng. Sci.*, 33, 493 (1978). Mill, C. C., *J. Oil Col. Chem. Assoc.*, 50, 396 (1967).
Brookfield viscometer	Aqueous suspensions, e.g., suspensions of paper clays, mineral pigments for paper coating, ceramics coatings, drilling muds, pharmaceuticals (i.e., slurries for injections, emulsion-based suspensions), process waste streams.	Applied for low-shear viscosity quality control of products. The device has already been described in this chapter. In slurry viscometry work, a suspension should be prepared at as high a solids content as possible — single-point viscosity measurement in the paper industry is normally made at 100 RPM. The sample is then diluted by 2% in solids and a second measurement is made. The root reciprocal viscosity is plotted vs. weight % solids and the amount of solids required for a viscosity of 5 poise is quoted. This is referred to as the *viscosity concentration or Clark viscosity*. This terminology is normally used in paper coating technology and for calcium carbonate slurry preparation.	TAPPI Standard: T 648-sm-54 Clark, N. D., *Trans. Br. Ceram. Soc.*, 49, 409 (1950). Beazley, K. M., *J. Coll. Int. Sci.*, 41(1), 105 (1972). Beazley, K. M., *TAPPI*, 50(3), 151 (1967).
Cone-and-plate		Already described above. One example of a commercial unit is the Ferranti-Shirley cone-and-plate instrument.	Huggenberger, L., W. Kogler, and M. Arnold, *TAPPI*, 62, 37 (1979). Borruso, D., A. Croce, and A. Seves, *Ind. della Carta*, 7, 100 (1969).

TABLE 3 (continued)
Summary of Major Viscometric Techniques Used Throughout Different Industries

Products	Type of viscometer	Operating principles — rangeability	Standard test methods	Suggested ref.
	Gallenkamp torsion viscometer	This instrument is largely applied in the U.K. for characterizing ceramic casting slips. It employs a small metal cylinder which is suspended from a torsion wire and dips into the test suspension. The position and deflection of the measurement head is recorded by a pointer and scale. The head is immersed in the test fluid, twisted through 360°, and released. The extent of overswing provides a measure of the fluidity of the sample.		Moore, F., *Rheology of Ceramic Systems*, Institute of Ceramics, Stoke on Trent, U.K., 1965.
	Vicat penetrometer	Often used for cement paste characterization, particularly in Europe and the U.K. Device uses a capped rod which moves vertically and carries a pointer over a scale at the same time. The lower end contains a needle which is used to penetrate the sample. The test sample is retained in a mold that sits on a nonporous plate. The needle is gently lowered into the sample and allowed to sink into it. The test is repeated several times until the needle no longer completely sink into the sample. The time from the first experiment (i.e., when the cement paste is first prepared with water) to the point of no penetration is measured and used as a record of the initial setting time.		Jones, T. E. R. and S. Taylor, *Sil. Ind.*, 4/5, 83 (1978).

REFERENCES

Andrews, R. D., N. Hofman-Bang, and A. V. Tobolsky, *J. Polymer Sci.,* 3, 669 (1948).
Aoki, Y. J., *Soc. Rheol. Jpn.,* 7, 20 (1979).
Bagley, E. B., *J. Appl. Phys.,* 28, 624 (1954).
Bair, S. and W. O. Winer, *Trans. ASME (F) J. Lubr. Technol.,* 101, 251 (1979).
Ballman, R. L., *Rheol. Acta,* 4, 137 (1965).
Barnes, H. A., in *Rheometry: Industrial Applications,* Walters, K., Ed., John Wiley & Sons, New York, 1980, pp. 31–118.
Bird, R. B., W. E. Stewart, and E. N. Lightfoot, *Transport Phenomena,* John Wiley & Sons, New York, 1967.
Brizitsky, V. I., G. V. Vinogradov, A. I. Isaev, and Yu Ya Podolsky, *J. Appl. Polymer Sci.,* 22, 665 (1978).
Calderbank, P. H. and M. B. Moo-Young, *Trans. Inst. Ch.E. (London),* 37, 26–33 (1959).
Chapman, F. M. and T. S. Lee, *SPIE J.,* 26, 37 (1970).
Chen, I. J. and D. C. Bogue, *Trans. Soc. Rheol.,* 16, 59 (1972).
Cheremisinoff, N. P., *Fluid Flow Pocket Handbook,* Gulf Publishing, Houston, 1984.
Cheremisinoff, N. P., *Instrumentation for Complex Fluid Flows,* Technomic Publishing, Lancaster, PA, 1986.
Cheremisinoff, N. P. and D. Azbel, *Fluid Mechanics and Unit Operations,* Ann Arbor Science Publishers, Ann Arbor, MI, 1983.
Cogswell, F. N., *Plastics Polymers,* 36, 109 (1968).
Cogswell, F. N., *J. Non-Newtonian Fluid Mech.,* 4, 9 (1978).
Cotten, G. R. and J. L. Theile, *Rubber Chem. Technol.,* 57, 749 (1979).
Dealy, J. M., paper presented at AICHE Meet., Miami Beach, FL, November 1978.
Dexter, F. D., J. C. Miller, and N. Philippoff, *Trans. Soc. Rheol.,* 5, 193 (1961).
Furuta, I., V. M. Lobe, and J. L. White, *J. Non-Newtonian Fluid Mech.,* 1, 207 (1976).
Goldstein, C., *Trans. Soc. Rheol.,* 18, 357 (1974).
Gortemaker, F. H., M. G. Hanson, B. de Cindio, H. M. Laun, and H. Janeschitz-Kriegel, *Rheol. Acta,* 15, 256 (1976).
Graessley, W. W., S. D. Glasscock, and R. L. Crawley, *Trans. Soc. Rheol.,* 14, 519 (1970).
Han, C. D., *J. Appl. Polymer Sci.,* 15, 2567, 2579, 2591 (1971).
Han, C. D., *Trans. Soc. Rheol.,* 18, 103 (1974).
Han, C. D., *Rheology in Polymer Processing,* Academic Press, New York (1976).
Han, C. D. and L. H. Drexel, *J. Appl. Polymer Sci.,* 17, 3429 (1973).
Ide, Y. and J. L. White, *J. Appl. Polymer Sci.,* 22. 1061 (1978).
King, R. G., *Rheol. Acta,* 5, 35 (1966).
Kreiger, I. M. and S. H. Maron, *J. Appl. Phys.,* 25, 72 (1954).
Lee, B. L. and J. L. White, *Trans. Soc. Rheol.,* 18, 467 (1974).
Lobe, V. M. and J. L. White, *Polymers Eng. Sci.,* 9, 617 (1979).
Lodge, A. S., *Trans. Faraday Soc.,* 52, 120 (1956).
Maxwell, B. and R. Chartoff, *Trans. Soc. Rheol.,* 9, 41 (1965).
Meissner, J., *Rheol. Acta,* 8, 78 (1969).
Meissner, J., *J. Appl. Polymer Sci.,* 16, 2877 (1972).
Metzner, A. B. and J. C. Reed, *AIChE J.,* 1, 434 (1955).
Middleman, S., *Trans. Soc. Rheol.,* 13, 123 (1969).
Mingawa, N. and J. L. White, *J. Appl. Polymer Sci.,* 20, 501 (1976).
Mocosko, C. W. and J. M. Lornsten, *SPIE Antec. Tech. Pap.,* 461 (1973).
Mooney, M. J., *Rheology,* 2, 231 (1931).
Munstedt, H., *Proc. 7th Int. Rheol. Congr.,* 496 (1976).
Nakajima, N. and E. R. Harrel, *Rubber Chem. Technol.,* 52, 9 (1979).
Novak, J. D. and W. O. Winer, *Trans. ASME (F) J. Lubr. Tech.,* 90, 580 (1968).
Philippoff, W., *Phys. Z.,* 35, 884 (1934).

Pliskin, I., *Rubber Chem. Technol.,* 46, 1218 (1973).

Skelland, A. H. P., *Non-Newtonian Flow and Heat Transfer,* John Wiley & Sons, New York, 1967.

Stevenson, J. F., *AICHE J.,* 18, 540 (1972).

Thomas, D. G., *Ind. Eng. Chem.,* 11(18), 55 (1963).

Tobolsky, A. V., *Structure and Properties of Polymers,* John Wiley & Sons, New York, 1960.

Treloar, L. R. G., *Trans. Faraday Soc.,* 36, 538 (1940); 37, 84 (1941); 43, 284 (1947).

Vinogradov, G. V., A. Ya Malkin, E. P. Plotnikova, O. Yu Sabsai, and N. E. Nikoilayeva, *Int. J. Polymer Materials,* 2, 1 (1972).

Vinogradov, G. V., B. V. Radushkevich, and V. D. Fikham, *J. Polymer Sci.,* A-2, 8 (1970).

Wales, J. L. S., *Rheol. Acta,* 8, 38 (1969).

Wales, J. L. S. and H. Janeschitz-Kriegel, *J. Polymer Sci.,* 5, 781 (1967).

Wales, J. L. S., J. L. den Otter, and H. Janeschitz-Kriegel, *Rheol. Acta,* 4, 146 (1965).

Walters, K., *Rheometry,* Chapman & Hall, London, 1975.

Weissenberg, K., *Proc. 1st Int. Rheol. Congr.* (1948).

White, J. L. and J. F. Roman, *J. Appl. Polymer Sci.,* 20, 1005 (1976).

Chapter 3

VISCOMETRIC TECHNIQUES AND ANALYSIS OF STEADY-STATE FLOWS

I. POISEUILLE FLOW AND THE CAPILLARY VISCOMETER

Constitutive equations are expressions that define the stress rate of the deformation relationship at every point in the fluid. In practice, it is relatively easy to measure macroscopic variables; for example, volume flow rate can be measured more easily than the velocity components at every point in the fluid. Consequently, a method of writing the functional dependence implied by a general constitutive equation is sought in terms of readily measurable macroscopic variables.

For a laminar steady-state flow through a circular tube or capillary of radius R, the volume flow rate Q is related to the velocity $v_z(r)$ by

$$Q = \int_0^R 2\pi r \, v_z(r) dr = \pi R^2 \langle V \rangle \tag{1}$$

where $\langle V \rangle$ denotes the mass average velocity. Integration by parts and application of the no-slip condition at the wall (i.e., $v_z = 0$ at $r = R$), yields:

$$Q = -\int_0^R \pi r^2 \frac{dv_z}{dr} \, dr \tag{2}$$

The only nonzero component of the rate of deformation tensor is $dv_z/dr = \Delta_{12}$. The shear rate-shear stress relationship can be expressed in the functional notation as

$$-\Delta_{12} = f(\tau_{12}) \quad \text{or} \quad \dot{\gamma} = f(\tau) \tag{3}$$

where τ is the shear stress component. From the dynamic equations for this flow it can be shown that the shear stress is linear across the tube radius [i.e., $\tau = C_o r/2 = (\Delta P/L)(r/2)$]. As described in Chapter 2, the wall shear stress is simply $\tau_w = R\Delta P/2L$.

By simple algebraic manipulation it can be shown that:

$$\frac{4Q}{\pi R^3} = \phi(\tau_w) = \frac{4}{\tau_w^3} \int_0^{\tau_w} \tau^2 \, f(\tau) \, d\tau \tag{4}$$

where $\phi(\tau_w)$ is defined by this expression.

Taking $d\phi/d\ \tau_w$ and using Leibniz' rule, one can develop an expression solving for $f(\tau_w)$, which is the well-known Weissenberg-Rabinowitsch-Mooney equation:

$$f(\tau_w) = \frac{3}{4}\ \phi + \frac{1}{4}\ \tau_w\ \frac{d\phi}{d\tau_w} \tag{5}$$

Because ϕ and τ_w are measurable in terms of macroscopic variables, Equation 5 allows one to find $f(\tau_w)$ over some range of τ_w. This functional dependence is identical with $f(\tau_w)$, however, so that with Equation 3, the shear behavior of the fluid is determined. The usual method is to plot ϕ vs. τ_w, and from this obtain $d\phi/d\tau_w$. Then the right side of Equation 4 is plotted against τ_w to give $f(\tau_w)$. From the resulting graph of $f(\tau_w)$ vs. τ, which is just Δ_{12} vs. τ_{12}, one can obtain and plot the viscosity η as a function of shear rate through the definition in Equation 9 in Chapter 2.

This analysis of flow in a capillary is based upon the assumption that simple shear flow exists. This is achieved in steady-state, laminar, isothermal flow in a tube of constant cross-section, as long as one does not consider regions near the entrance and exit of the tube. In these "end" regions the flow is changing from (or to) its previous (or future) distribution outside the tube. The length of an "end" region is generally a function of tube diameter and some dynamic parameters. For example, for a Newtonian fluid, the "entrance length", the length of tube required to achieve the fully developed simple shear flow, depends upon the Reynold's number.

One method of minimizing the effect of the entrance length is to use a viscometer tube so long that the pressure drop over the entrance region is very small compared to the drop over the entire tube. Generally, this means that if L is the total tube length, then L_e/L must be small, perhaps on the order of 0.01. Often this restriction may demand tubes too long for practical purposes.

A method for eliminating the influence of end effects exists which uses two tubes identical except in length. The shorter tube must be longer than the entrance length, so that both tubes have a finite region in which the flow is in simple shear. Consider two such tubes attached to the bottoms of identical pressurized reservoirs, as illustrated in Figure 1. Suppose that the pressure drops across each tube are adjusted so that the volume flow rates are identical and the reservoirs contain identical fluids. Let ΔP_L and ΔP_S be the total pressure drops measured across the long and short tubes, respectively. If the volume flow rate is the same in both tubes, one would then expect the entrance and exit lengths and the pressure drops across these lengths to be identical in both tubes. Then

$$\Delta P_L = \Delta P_{en} + \Delta P_{ex} + \Delta P_{fdL} \tag{6}$$

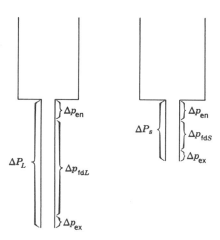

FIGURE 1. Definition sketch for the treatment of capillary flow data.

$$\Delta P_S = \Delta P_{en} + \Delta P_{ex} + \Delta P_{fdS} \qquad (7)$$

where ΔP_{en} and ΔP_{ex} are the pressure drops across the entrance and exit regions and ΔP_{fdL} is the pressure drop across the long (L) or short (S) tube that arises strictly from the fully developed flow. Since the flow rates are the same, the pressure gradients in the fully developed sections are identical, and

$$\Delta P_{fdL}/L_L = \Delta P_{fdS}/L_S \qquad (8)$$

where L_L and L_S are the lengths over which the flow is fully developed in the large and small tubes. Combining the last three expressions, we obtain:

$$\frac{\Delta P_{fdL}}{L_L} = \frac{\Delta P_L - \Delta P_S}{L_L - L_S} \qquad (9)$$

This is the pressure gradient free of entrance and exit effects, although either viscometer tube may have involved appreciable end effects. It should be noted that the tube lengths should be sufficiently different so that the pressure difference $\Delta P_L - \Delta P_S$ is not subject to large error.

The shear stress can be expressed by a power law model:

$$\tau_w = K'\phi^n \qquad (10)$$

and from Equation 4 one can write

$$\tau = \left(\frac{4n}{3n + 1}\dot{\gamma}\right)^n K' \qquad (11)$$

This expression can be stated as the standard form for the power law

$$K = K' \left(\frac{4n}{3n + 1} \right)^n \tag{12}$$

II. VISCOUS HEAT GENERATION

The generation of heat through viscous dissipation can lead to significant temperature variations across the shear fields. Because fluid properties such as viscosity are strongly temperature dependent the shear stress-shear rate relation is considerably altered by non-isothermal effects. Hence, a shear stress-shear rate curve obtained under nonisothermal conditions does not reflect the basic fluid response independently of any temperature-dependent effects. The challenge then is to take data subject to viscous heating effects and separate that part of the response due to nonisothermal behavior from that part of the response due to non-Newtonian behavior.

Analyses reviewed thus far are applicable to nonisothermal flows; however, these analyses do not allow a separation of the temperature-dependent behavior from the overall response of the fluid. This can be done only if a model for the fluid (or constitutive equation) is assumed, and if the temperature variation of the fluid parameters is specified. This results in a parameter fitting procedure in which data obtained at different shear rates and temperatures are manipulated until a consistent set of fluid constants that are independent of shear rate and temperature is obtained.

In Poiseuille flow, if heating is important, one must solve the energy equation for laminar flow in cylindrical coordinates. With appropriate boundary conditions the problem to be solved is the so-called energy equation:

$$\rho \, \hat{C}_P \, v_z \, \frac{\partial T}{\partial z} = \frac{1}{r} \frac{\partial}{\partial r} \left(rk \, \frac{\partial T}{\partial r} \right) + \Phi \tag{13}$$

$$T = T_0 \text{ at } z \leqslant 0$$

$$\frac{\partial T}{\partial r} = 0 \text{ at } r = 0$$

$$\left. \begin{array}{c} -k \, \dfrac{\partial T}{\partial r} = q \\[2mm] \text{or} \\[2mm] T = T_0 \end{array} \right\} \quad \text{at } r = R$$

where T_0 is the reservoir temperature of the fluid, and k, ρ, and $\hat{C}_p$ are the fluid thermal conductivity, density, and heat capacity, respectively. The third boundary condition is arbitrary, in that it depends upon actual operating conditions. Typical conditions used state that the heat flux across the tube wall may be a constant q (including zero, the adiabatic condition), or the temperature of the wall may be specified as a constant.

Equation 13 is approximate in that small radial velocities arising from the effect of viscosity variation on the flow field are neglected, so that terms such as $v_r \, \partial T/\partial r$ do not appear. In addition, it is assumed that axial conduction of heat is unimportant, so that the term $\partial(k\partial T/\partial z)/\partial z$ does not appear. The fluid is assumed incompressible, and ρ is taken as a constant. These assumptions are valid for moderate temperature variations.

If the velocity profile is assumed to be perturbed by the varying temperature field, then the energy equation must be solved in conjunction with the dynamic equation for this flow, given by

$$0 = -\frac{\partial P}{\partial z} + \frac{1}{r}\frac{\partial}{\partial r}(r\tau) \tag{14}$$

subject to the boundary conditions

$$v_z = 0 \quad \text{at r} = R$$

$$\partial v_z/\partial r = 0 \quad \text{at r} = 0$$

$$P = P_1 \quad \text{at z} = 0$$

Again, radial velocities are ignored.

If a model relating τ to $\dot{\gamma}$ is introduced, and if a functional form for the temperature dependence of the parameters of this model is introduced, the resulting set of equations constitutes a nonlinear coupled boundary value problem which cannot be analytically solved.

If fluid properties are assumed to be constant, in which case the isothermal velocity profile is known, solutions exist from which the temperature profile may be calculated as a function of axial position. Bird[1] gives solutions for the case of the power law fluid, with either the isothermal or adiabatic wall. Siegel et al.[2] give solutions for the Newtonian fluid with boundary conditions specifying constant prescribed flux at the tube wall. The simplest case is that of Brinkman[3] for the Newtonian fluid with temperature-independent properties, with either isothermal or adiabatic wall. None of the studies accounts for the alteration of the flowfield arising from viscosity variation across the radius of the capillary. As such they may not be used to correct viscometry data. Their primary use is in the estimation of the temperature rise to be expected. If the estimate of the maximum temperature rise is small, say $<1°C$,

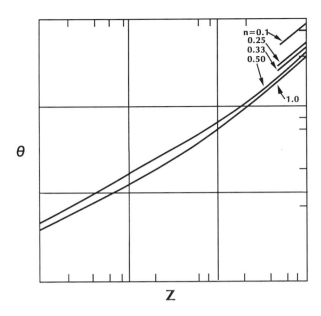

FIGURE 2. Dimensionless wall temperature rise down the length of a capillary, due to viscous heating.

then one might judge that the results need not be corrected at all for heating effects. If the estimated temperature rise is large, then one is faced with the problem of rejecting the data or accepting the results as subject to significant error.

In experiments with ethylene propylene diene monomer (EPDM) rubbers the wall of a capillary behaves in a nearly adiabatic manner under normal operating conditions. Hence, estimates of temperature rise based upon the adiabatic theories are useful. Figure 2 shows Bird's solutions for the power law fluid, plotted as a dimensionless wall temperature which is a function of a dimensionless axial variable, with n as a parameter. The dimensionless variables are defined as

$$\Theta = (T - T_0) \frac{4k \ K^{1/n}}{D^2 \ \tau_w^{(n+1)/n}} \left(\frac{3n \ + \ 1}{2n}\right)^2 \tag{15}$$

and

$$Z = \frac{4}{Pe} \frac{n \ + \ 1}{3n \ + \ 1} \frac{Z}{D} \tag{16}$$

τ_w is the wall shear stress and Pe is the Peclet number ($= D \langle V \rangle \rho \hat{C}_P / k$).

It is interesting to examine the manner in which variables such as $\langle V \rangle$ and D affect the maximum temperature rise. This is shown simply by rewriting Equation 15 as follows:

$$T - T_0 = \frac{D^2K}{4k} \dot{\gamma}^{n+1} \left(\frac{2n}{3n+1}\right)^2 \Theta\left(n, \frac{z}{D}, Pe\right)$$ (17)

where τ_w has been replaced by $K\dot{\gamma}^n$. If interest is at conditions under high shear rates, then

$$\dot{\gamma} = \frac{3n+1}{4n} \left(\frac{8\langle V \rangle}{D}\right)$$

and thus a choice must be made between increasing $\dot{\gamma}$ by increasing $\langle V \rangle$ or by decreasing D. As an example, if we consider $4 \times \dot{\gamma}$ then the coefficient of Θ is decreased by a factor of $(1/4)^2(4)^{n+1}$. For $n = 1$ the factor is unchanged, while for $n = 1/2$ the factor is halved. On the other hand, if $\dot{\gamma}$ is increased by increasing $\langle V \rangle$ by 4, at constant D, then the coefficient of Θ increases directly as $\dot{\gamma}^{n+1} = (4)^{n+1}$. For $n = 1$ and $1/2$ this factor is 16 and 8, respectively. It becomes clear that the heating effect is minimized if $\dot{\gamma}$ is increased by using smaller diameter capillaries. In fact Θ is a function of D and $\langle V \rangle$ by way of the Peclet number. However, the Peclet number dependency is weak under usual experimental conditions, and hence the above observations apply. Of course, it has been assumed that K and n are unchanged by an increase in $\dot{\gamma}$. While this is not rigorous, in the sense that no fluid truly possesses power law behavior, the approximation does not appear to be bad.

Capillary experiments involving highly viscous materials should always be accompanied by an estimate of temperature rise due to viscous heating. In general, many applications of viscosity data simply do not warrant the effort required to ascertain and correct for nonisothermal effects. On the other hand, one can find studies in the literature that draw conclusions based on the assumption that the high shear rate viscosity of highly viscous polymer melts is free of viscous heating perturbations.

Middleman[4] provides a nomograph for estimation of temperature rise due to viscous dissipation. The nomograph is based upon the solution for fully developed capillary flow of a power law fluid. This solution provides the temperature rise at the wall of the capillary, at some axial position z, under the assumption that the wall is adiabatic. To simplify the nomograph, approximations were made which essentially remove any dependence on the power law index n. In examples tested, the nomograph yields a temperature rise within a factor of 2 of the analytical solution, and hence, provides a rapid estimate of the order of magnitude of the effect of viscous dissipation in a highly viscous capillary flow. All scales are in centimeters per gram per second units. Conversion factors are given below. A short table of thermal properties of common polymers is also given. Note especially that k must be in centimeters per gram per second units. To use the nomograph, move from left to right across the chart. For a given set of data, begin by connecting points on the $\dot{\gamma}$ and τ scales, and locate the intersection of this line with reference scale 1. Connect that point, through D, to reference scale 2, and so on across the other scales to the temperature scale. The value of Θ is an estimate of the temperature rise at the wall of the capillary, at the given value of z/D.

The following are typical literature values of thermal properties reported by Middleman:

	$\left(\dfrac{\text{g/cm}}{\text{s}^3/°\text{C}}\right)$	$k/\rho\hat{C}_p\left(\dfrac{\text{cm}^2}{\text{s}}\right)$
Benzene	1.5×10^4	8.5×10^{-4}
Water	6.3×10^4	1.5×10^{-3}
Polybutadiene/styrene copolymer	2.1×10^4	1.2×10^{-3}
Polybutene	1.2×10^4	7.4×10^{-4}
Polyethylene	3.3×10^4	1.3×10^{-3}
Polyethylene tere-phthalate	1.2×10^4	5.1×10^{-4}
Polystyrene	1.2×10^4	6.2×10^{-4}

Conversion Factors

To obtain k in g/cm/s³ °C, multiply k in Cal/s/cm/°C by 4.2×10^7; k in BTU/h/ft/°F by 1.7×10^5.

To obtain k/ρCp in cm²/sec, multiply k/ρĈp in ft²/hr by 0.26.

To obtain $k/\rho C_p$ in g/cm/s² (dyne/cm²), multiply in lb · ft$_f^2$ by 480; in lb · in.$_f^2$ by 6.9×10^4.

To obtain D in centimeters, multiply D in in. by 2.54; D in mil by 2.54×10^{-3}.

Note: It should suffice to use nominal values for $\dot{\gamma}$ and τ, i.e., $\dot{\gamma} = 8 \langle V \rangle/D$ and $\tau = D\Delta P/4L$. For polymer melts, $\dot{\gamma}$ would be underestimated but τ would be overestimated (assuming no corrections are made to ΔP) and the errors would tend to be compensatory.

We now direct attention to the case of Couette flow. To assess viscous dissipation one must solve the dynamic and energy equations together:

$$0 = \frac{d}{dr}(r^2\tau) \tag{18}$$

$$0 = \frac{1}{r}\frac{d}{dr}\left(rk\frac{dT}{dr}\right) + \Phi \tag{19}$$

subject to boundary conditions

$$v_\theta = 0 \quad \text{and} \quad T = T_0 \quad \text{at} \quad r = R_0$$

$$\left.\begin{array}{l} v_\theta = V = 2\pi\Omega R \quad \text{and } T = T_0 \\ \qquad\qquad \text{or } \dfrac{\partial T}{\partial r} = 0 \end{array}\right\} \quad \text{at } r = R$$

Note T and v_θ are functions only of radial position, and hence may be described by ordinary differential equations. Because of this it is possible to introduce some realistic complications into the analysis which cannot be readily handled in capillary flows that are described by partial differential equations.

The technique outlined below follows that used by Turian and Bird.[5] It is necessary to assume a constitutive equation and a temperature dependence of fluid properties. The power law is a useful model for the rheological behavior; the fluid properties are taken to be functions of temperature expressible in a power series about some reference temperature at which properties are assumed known. Shear stress is defined by:

$$\tau = K\,\dot{\gamma}^n \tag{20}$$

and K is taken as an expansion formula

$$K = K_0[1 - K_1(T - T_0) + K_2(T - T_0)^2 - \ldots] \tag{21}$$

For moderate temperature changes the thermal conductivity may be taken as constant.

Equations 18 and 19 may be solved by a perturbation method. The results outlined below depend on whether one assumes an isothermal inner wall or an adiabatic inner wall. For an isothermal wall, the maximum temperature rise is

$$\begin{aligned} \frac{\theta_{max}}{B_R} &= \frac{(T - T_0)_{max}}{T_0 B_R} \\ &= \frac{n^2 C_0^{1 + (1/n)}}{4}\left[1 - \frac{n}{2}f(s,n)\left(1 - \ln\frac{n}{2}f(s,n)\right)\right] \end{aligned} \tag{22}$$

The shear rate at the moving wall is

$$\dot{\gamma}_R = \dot{\gamma}_0 \left\{ 1 - \frac{\beta B_R sn^2 C_0^{1+(2/n)}}{16} \left[1 - S^{-4/n} - n(1 - S^{-2/n}) f(3,n) \right] \right\} \quad (23)$$

For the case of an adiabatic wall, the maximum temperature rise is

$$\frac{\theta_{max}}{B_R} = \frac{n^2 C_0^{1+(1/n)}}{4} \left[1 - S^{-2/n} - \frac{2}{n} S^{-2/n} \ln S \right] \quad (24)$$

The shear rate at the moving wall is

$$\dot{\gamma}_R = \dot{\gamma}_0 \left\{ 1 - \frac{\beta B_R sn^2 C_0^{1+(2/n)}}{16} \left[1 - S^{-4/n} - n(1 - S^{-2/n}) f(S,n) \right] \right\} \quad (23)$$

In the above relationships, $f(s,n) = (1 - S^{-2/n})/\ln S$, $C_0 = S^{2-n} (\gamma_0/\Omega)^n$ and B_R is the dimensionless Brinkman number $\left(B_R = \frac{R^{1-n} K_0 V^{1+n}}{KT_0} \right)$. Finally, $\beta = K_1 T_0$ is a measure of the temperature sensitivity of K. If βB_R is not small compared to unity, then the solutions above are inaccurate. In that case more terms in the perturbation series are required.

Expressions for $\dot{\gamma}_R$ are in the form of a correction factor multiplied by the isothermal shear rate. To apply the correction factor it is necessary to have at hand values of fluid properties such as β, K_0, and n. This poses a dilema (i.e., these values are not known *a priori*). This suggests that correction for nonisothermal effects must be based on a trial and error technique. The data are treated as if free of heating effects, and K_0 and n are estimated. The maximum temperature may then be estimated and a second set of data obtained at this temperature level. From estimates of K_0 at two temperatures an estimate of β may be obtained. From β, K_0, and n an approximate correction factor is obtained and new values of β, K_0, and n are calculated. The procedure is repeated until satisfactory convergence is obtained.

REFERENCES

1. **Bird, R. B., Stewart, W. E., and Lightfoot, E. N.,** *Transport Phenomenon,* John Wiley & Sons, New York, 1967.
2. **Siegel, R., Sparrow, E. M., and Hallman, T. M.,** *Appl. Sci. Res.,* A7, 386, 1958.
3. **Brinkman, H. C.,** *Appl. Sci. Res.,* A2, 120, 1951.
4. **Middleman, S.,** *The Flow of High Polymers: Continuum and Molecular Rheology,* Interscience, New York, 1968.
5. **Turian, R. M. and Bird, R. B.,** *Chem. Eng. Sci.,* 18, 689, 1963.

Chapter 4

NORMAL STRESS MEASUREMENTS

I. INTRODUCTION

Some of the techniques of viscometry were discussed in previous chapters. Many fluids, when subjected to a simple shear flow, develop not only shear stresses but also normal stresses. These normal stresses manifest themselves in the so-called "normal stress phenomena" which commonly occur in viscometric experiments. These stresses also materialize in many important flows. For example, when a rod (such as the shaft of a stirrer) is rotated about its axis perpendicular to the free surface of a Newtonian liquid, the liquid surface is depressed in the neighborhood of the rod, as a consequence of centrifugal forces that accompany the induced rotational flow. Some fluids, however, are observed to develop an elevated surface at the rod. This "rod climbing" effect is perhaps the best known normal stress phenomenon, and is called the "Weissenberg effect".

When certain non-Newtonian fluids (e.g., polymer melts) are ejected from an orifice or tube, the resulting jet is commonly observed to expand to a diameter much larger than its initial ejection diameter. This effect is often referred to industrially as "die swell" or "extrudate swelling".

This chapter is concerned with information of fundamental rheological significance from the measurement of normal stress phenomena. In contrast to the measurement of shear stresses, the measurement of normal stresses includes the isotropic stress $p\delta$ in addition to the dynamic stress τ. In order to generate information specifically about the dynamic stress components, it is necessary to "remove" the influence of p from the measurements. This is most commonly done by expressing results in terms of stress differences since, for example

$$T_{11} - T_{22} = -p + \tau_{11} - (-p + \tau_{22}) = \tau_{11} - \tau_{22} \qquad (1)$$

and the isotropic component does not appear. When this is not possible it may be necessary to impose some assumptions upon p.

The fundamental meaning of the isotropic component p is still not clear for a viscoelastic material. While the classical definition is

$$p = p_m = -\frac{1}{3}(T_{11} + T_{22} + T_{33}) \qquad (2)$$

for such fluids, so that p is by definition "mean normal stress", it is not obvious that this is rigorous.

As in the case of viscometry, the flows studied approximate simple shear flows. In contrast to the techniques of viscometry, many of the methods used for normal stress measurement are incompletely developed, in the sense that the effects of perturbations which cause deviations from a simple shear flow, such as exit effects in capillary flows, are not fully understood. Techniques for removing the influence of such effects from data are subject to assumptions not yet theoretically justified.

II. ANALYSIS OF THE CAPILLARY JET

When a fluid discharges from a tube as a free jet it exerts an axial thrust upon the tube. This thrust can be calculated as the difference between the flux of axial momentum leaving the tube and the axial stresses that exist in the fluid as it is ejected. The thrust F can be expressed as follows:

$$F = \int_0^R \rho \, v_z^2 \, 2\pi \, r \, dr \, - \, \int_0^R T_{zz} \, 2\pi \, r \, dr \tag{3}$$

This expression is subject to the assumption that surface tension and gravity do not contribute significantly to the axial momentum.

It is possible to invert this relationship via a technique very similar to that used in developing the Weissenberg-Rabinowitsch-Mooney equation and thereby solve for T_{zz} at the tube wall. The resultant expression is

$$(T_{zz})_R = \rho \langle V \rangle^2 \left[\frac{3n + 1}{n} - 2 \int_0^1 \left(\frac{v_z}{\langle V \rangle} \right) \frac{r}{R} \, d\left(\frac{r}{R} \right) \right]$$

$$- \frac{F}{\pi R^2} \left(1 + \frac{1}{2n} \frac{d \ln F}{d \ln 8 \langle V \rangle / D} \right) \tag{4}$$

The shear rate at the tube wall can be obtained from the Weissenberg-Rabinowitsch-Mooney expression. It is therefore possible to obtain T_{zz} and the corresponding shear rate. This result is subject to the assumption that the velocity and stress distributions at the tube exit correspond to those for fully developed flow. Recall that n is the slope of the log τ vs. log $\dot{\gamma}$ flow curve at the conditions of interest.

Introducing the components of τ in place of the components of T,

$$(T_{zz})_R = -p(R,L) + (\tau_{zz})_R \tag{5}$$

$(T_{zz})_R$ is measured, using Equation 4, at the exit $(Z = L)$. Hence, the isotropic stress component $p(R,L)$ at the tube exit must be known in order to find $(\tau_{zz})_R$.

It can be assumed that p(R,L) denotes atmospheric pressure at the tube exit. Examination of the dynamic equations, however, shows this to be false. The radial component of the equations of motion gives

$$0 = -\frac{\partial p}{\partial r} + \frac{\partial \tau_{rr}}{\partial r} + \frac{\tau_{rr} - \tau_{\theta\theta}}{r} \tag{6}$$

Integration from $r = 0$ to some arbitrary r gives

$$p(r,z) = p(0,z) + \tau_{rr} + \int_0^1 \frac{\tau_{rr} - \tau_{\theta\theta}}{r} \, dr \tag{7}$$

This is subject to the reasonable assumption that $\tau_{rr} = 0$ along the center line of the tube, where the deformation rate vanishes. Hence,

$$(T_{11})_R = -p(0,L) + (\tau_{11} - \tau_{22})_R - \int_0^R \frac{\tau_{22} - \tau_{33}}{r} \, dr \tag{8}$$

In the absence of dynamic normal stresses it follows that the only isotropic stress would be ambient pressure P_o and $p(0,L) = p(R,L) = P_o$ would be true.

The axial stress T_{11} may also be determined from a jet expansion experiment. Defining the final radius and velocity of the jet as R_j and V_j and assuming that the liquid is free of stress (relative to the ambient pressure) at this point, the flux of momentum across any position beyond the region where these final values are achieved is $\pi \rho R_j^2 V_j^2$. Because momentum is conserved in the jet, however, this quantity must just equal the thrust exerted on the tube. Hence, in Equation 4, one can replace F by $\pi \rho R_j^2 V_j^2$

$$(T_{11})_R = \rho\langle V\rangle^2 \left[\frac{3n+1}{n} - 2\int_0^1 \left(\frac{v}{\langle V\rangle}\right)^2 \frac{r}{R} \, d\left(\frac{r}{R}\right) \right.$$

$$\left. - \frac{1}{n\alpha^2}\left(1 + n - \frac{d \ln \alpha}{d \ln 8\langle V\rangle/D}\right) \right] \tag{9}$$

where $\alpha = R_j/R$. Equation 8 is applicable regardless of whether one measures thrust or expansion.

The stress T_{11} is associated with the normal stress in equilibrium with the fully developed shear flow within the capillary. There is, however, a viscous contribution to T_{11} which arises from the relaxation of the velocity profile subsequent to ejection of the liquid as a free jet. In Newtonian fluids, for which there is no normal stress except that developed by profile relaxation, this viscous stress can be large enough to cause the jet to swell. For Reynolds numbers larger than about 200, however, the viscous contribution is negligible.

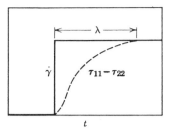

FIGURE 1. Concept of a relaxation time, based on the hypothetical response of a viscoelastic fluid to a step change in shear rate.

In non-Newtonian fluids it is not possible to separate the viscous effect from the primary normal stress effect. It is usually assumed that the viscous stresses are negligible under the experimental conditions which are peculiar to a study of jets of polymeric fluids. The validity of this assumption can be checked by plotting values of T_{11} obtained with different diameter tubes against shear rate. Since the viscous effect would not be a function of shear rate, but rather would be a function of a viscous parameter, such as a non-Newtonian Reynolds number, data taken in different diameter tubes would fall on a single curve, when plotted as T_{11} vs. $\dot{\gamma}$ only if free of viscous effects.

At very high flow rates the ratio R_j/R approaches a limiting value which depends only upon the slope n of the log τ vs. log $\dot{\gamma}$ curve for the particular fluid. For the power law fluid this limiting value is given by

$$\alpha_\infty = R_j/R = \sqrt{(2n + 1)/(3n - 1)} \tag{10}$$

and lies in the range $0.866 \leq \alpha \leq 1$ for $1 \geq n \rangle 0$.

A characteristic relaxation time for a fluid can be defined as the time required for the stress in the fluid to adjust to a step change in shear rate. Figure 1 gives a possible quantitative definition for a relaxation time. One would expect that the normal stresses are in equilibrium with the shear field in the tube if the residence time $L/\langle V \rangle$ is larger than the relaxation time λ. The criterion for fully developed normal stresses is

$$\frac{L_n/\langle V \rangle}{\lambda} = k_n \quad \text{or} \quad \frac{L_n}{D} = k_n \frac{\langle V \rangle \lambda}{D} \tag{11}$$

The parameter k_n must be established through experiment (but one would expect it to be of the order of unity). Estimates of λ are on the order of magnitude of 10^{-2} s for fluids that show significant expansion. For example, typical values of $\langle V \rangle$ and D are $\langle V \rangle = 500$ cm/s and $D = 0.1$ cm; a typical estimate for L_n/D is 50.

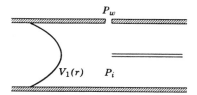

FIGURE 2. Definition sketch for normal stress measurement using a pitot tube in laminar Poiseuille flow.

Since shear rates in capillary flows are generally high (around 10^4/s), and since fluids that exhibit significant swelling are fairly viscous, heat generation within the capillary can be appreciable. In addition to producing a temperature different than the one that the fluid is presumed to have on the basis of the reservoir temperature, the temperature field alters the velocity profile. The velocity profile appears in Equation 4 and it is clear that viscous heating effects could complicate the proper interpretation of normal stress measurement.

Under normal conditions jet experiments involve shear rates in the range 10^{-4} to 10^5/s. To extend the range over which useful normal stress data may be obtained in capillary flow, a pitot tube method is suitable to lower shear rates. Using Figure 2 as a reference, the pressure P_i would be an "impact" pressure, involving the axial normal stress in addition to the dynamic pressure $\frac{1}{2} \rho v_1^2(r)$ or

$$P_i = -T_{11}(r) + \frac{1}{2} \rho v_1^2 (r) \tag{12}$$

The wall pressure P_w would be just $-T_{22}(R)$.

From the dynamic equations for laminar Poiseuille flow, one finds for the radial direction

$$T_{22}(R) = -p(0,z) - \int_0^R (\tau_{22} - \tau_{33}) \frac{dr}{r} \tag{13}$$

and, for the axial direction, Equation 8 with R and L replaced by r and z we obtain:

$$T_{11}(r) = -p(0,z) + (\tau_{11} - \tau_{22}) - \int_0^r (\tau_{22} - \tau_{33}) \frac{dr}{r} \tag{14}$$

The pressure difference $P_i - P_w$ becomes:

$$P_i - P_w = -(\tau_{11} - \tau_{22}) + \frac{1}{2} \rho v_1^2(r) - \int_r^R (\tau_{22} - \tau_{33}) \frac{dr}{r} \tag{15}$$

Experiments shows that the normal stress difference $\tau_{22} - \tau_{33}$ is considerably smaller than the normal stress difference $\tau_{11} - \tau_{22}$. It is therefore normal practice to assume that $P_i - P_w$ allows the determination of the normal stress difference $\tau_{11} - \tau_{22}$. This assumes that the velocity profile is known. If, however, viscous heating is significant the velocity profile will be perturbed, and the calculation of $\tau_{11} - \tau_{22}$ will be in error. At low shear rates, viscous heating effects are likely to be minimal. A thermocouple attached to the pitot tube would serve to indicate the presence of a significant temperature variation across the capillary.

Note also that Equation 15 indicates that a pitot tube may not be used directly to measure the velocity profile in a fluid that exhibits normal stress phenomena. Any pitot tube measurement is subject to the criticism that the probe distorts the flow, and hence does not measure the velocity profile which would exist if the probe were not present. The success of pitot tube measurements in Newtonian flows lends some confidence to its use in more complex flows. However, with a viscoelastic flow in the neighborhood of an obstruction, the flow may be different from a Newtonian flow which is kinematically similar at some distance from the obstruction. The difference arises primarily from a markedly different behavior of the boundary layer in viscoelastic stagnation flows. Evidence based on markedly reduced heat transfer coefficients from heated cylinders placed in a stream of a viscoelastic fluid supports this view.

III. NORMAL STRESSES IN THE CONE-AND-PLATE CONFIGURATION

In discussing the cone-and-plate system as a viscometer it was stated that the velocity field is

$$v_\phi = v_1 = r \, \Omega \, \psi/\psi_0 \tag{16}$$

and the shear rate is $\dot\gamma = \Omega/\psi_0$. The r-component of the dynamic equations, in spherical coordinates, is reducible to the form

$$-\rho \, \frac{v_1^2}{r} = -\frac{\partial P}{\partial r} + \frac{\partial \tau_{33}}{\partial r} - \frac{\tau_{11} + \tau_{22} - 2\tau_{33}}{r} \tag{17}$$

Since shear rate is independent of r, and since τ is a function only of shear rate, terms such as $\partial\tau_{33}/\partial r$ are identically zero. The stress measured by pressure taps in the plate ($\psi = 0$) is

$$T_{22}(r,0) = -p(r,0) + \tau_{22} \tag{18}$$

If T_{22} is introduced into the dynamic equation, the result is

$$\frac{\partial T_{22}}{\partial r} = -\left(\frac{\rho\, v_1^2}{r} - \frac{X(\dot{\gamma})}{r}\right) \tag{19}$$

where $X = \tau_{11} + \tau_{22} - 2\tau_{33}$. If the isotropic component of T is taken as the mean normal stress, then $X = -3\tau_{33}$. If Equation 19 holds at the plate surface ($\psi = 0$), then the measured stress distribution becomes:

$$\frac{\partial T_{22}}{\partial r} = -\frac{3\tau_{33}}{r} \quad \text{or} \quad \frac{\partial T_{22}}{\partial \ln r} = -3\tau_{33} = \text{Constant} \tag{20}$$

Deviations between this prediction and experimental results may be ascribed to inertial effects that cause the velocity profile to deviate from the simple shear expression assumed above. A first-order estimate of inertial effects is possible for the Newtonian fluid if the simple shear velocity profile is inserted in Equation 17 so that

$$\frac{1}{\rho}\frac{\partial P}{\partial r} = \frac{v_1^2}{r} = r\,\Omega^2\,(\psi/\psi_0)^2 \tag{21}$$

Integration yields

$$(P - P_0) = \frac{1}{2} P\,\Omega^2\,(\psi/\psi_0)^2\, r^2 + \rho C_N(\psi) \tag{22}$$

where P_0 is some arbitrary pressure and C_N is a function of ψ, but not r. For $\psi = 0$, the pressure along the plate is independent of radial position, according to this result.

Experimental measurements are not in agreement with this prediction; instead they indicate a variation of p with r. A possible interpretation of this result seems to have no physical foundation, but does lead to a prediction consistent with the experiment. It is claimed that the stress measured at the plate is really the ψ-average of the stress distribution between the cone and plate.

A detailed evaluation of the dynamic equations for a non-Newtonian fluid, coupled with the assumption that the stress measured along the plate surface, gives the result

$$\overline{T}_{22} + P_0 - \tau_{22} = -3\tau_{33}\left(\frac{1}{3} - \ln\frac{R}{r}\right) - \frac{1}{6}\rho\,\Omega\,R^2\left(\frac{r^2}{R^2} - \frac{3}{5}\right) \tag{23}$$

τ_{22} is not averaged because it is a function only of shear rate and so is independent of ψ, and P_0 is taken to be the ambient pressure.

A plot of the measured quantity

$$\overline{T}_{22} + P_O + \frac{1}{6} \rho \, \Omega \, R^2 \left(\frac{r^2}{R^2} - \frac{3}{5} \right)$$

against $1n (R/r)$ should be linear, with a slope equal to $3\tau_{33}$ and a zero intercept at a radial position r_o such that

$$\ln(R/r_O) = \frac{1}{3}(1 - \tau_{22}/\tau_{33}) \tag{24}$$

Hence, both τ_{33} and τ_{22} may be obtained directly from such an experiment, if the inertial correction is known.

The inertial correction may be determined experimentally by measuring the stress distribution along the plate for a Newtonian fluid. If inertial effects perturb the stress field without causing a significant deviation of the velocity field from that assumed in Equation 21, then a non-Newtonian fluid will behave, with respect to inertial forces, like a Newtonian fluid, as the shear field will be uniform throughout the fluid. Thus, the inertial correction at a particular shear rate can be obtained from the measured stress distribution of a Newtonian fluid whose viscosity is the same as the apparent viscosity of the non-Newtonian fluid at that shear rate.

Because the inertial stresses depend upon Ω, but not upon ψ_0, data taken with different cone angles should be a function only of Ω/ψ_0 when the proper inertial corrections have been made. Thus, one should always take data with more than a single cone angle whenever possible.

If the total vertical force exerted against the cone is measured, it is possible to determine the normal stress difference T_{22} directly, without a measurement of the stress distribution in the system. For the small cone angles usually employed, this force may be calculated from

$$F = -\pi R^2 \left[(-p(R) + \tau_{33}) + \frac{1}{2}(\tau_{22} - \tau_{11}) \right] \tag{25}$$

If the system is in equilibrium with the atmosphere on its outer boundary, then

$$-p(R) + \tau_{33} = T_{33}(R) = 0$$

and it follows that

$$\tau_{11} - \tau_{22} = 2F/\pi R^2 \tag{26}$$

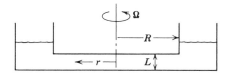

FIGURE 3. Definition sketch for parallel plate torsional shear flow.

Inertial effects may be introduced as before. The most common procedure is to correct F on the basis of experiments with Newtonian fluids of appropriate viscosities. Then $\tau_{11} - \tau_{22}$ is calculated from F', a corrected force from which the inertial effect has been removed.

IV. PARALLEL PLATE-CONFIGURATION

Another system useful for the measurement of normal stresses at low shear rates is the parallel-plate torsional configuration, most easily achieved in the space between the bottoms of two concentric cups. Figure 3 shows the geometry for this system.

If inertial effects are unimportant, and if edge effects at the radial boundary can be ignored, then the velocity profile may be shown to be

$$v_1 = \Omega \, rz/L \tag{27}$$

The only nonzero components of the rate of deformation tensor are

$$\dot{\gamma} = \Delta_{12} = \Delta_{21} = \Omega \, r/L \tag{28}$$

The shear rate varies radially in this system.

The governing expressions for this system is

$$T_{22}(r) + p(0) = \tau_{22} - \tau_{33} + \int_0^r \frac{\tau_{11} - \tau_{33}}{r} \, dr \tag{29}$$

subject to the assumption that T_{22} and T_{33} are zero at $r = 0$, as the shear rate vanishes at that point.

Because T_{22} may be measured as a function of r, the right side of Equation 29 may be obtained. However, further information about the individual stress differences requires arbitrary assumptions about their relative values or their dependencies on shear rate.

A more useful result is obtained from the determination of the total vertical force exerted against the upper plate. The force F is given by

$$F = -2\pi \int_0^R T_{22}(r)r \, dr \tag{30}$$

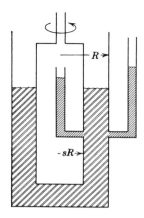

FIGURE 4. Definition sketch for normal stress determination in cylindrical Couette flow.

If $\dot{\gamma}$ is substituted for r through Equation 28 and F is differentiated with respect to $\dot{\gamma}_R$ the result is

$$-3\tau_{22}(\dot{\gamma}_R) = \frac{2F}{\pi R^2}\left(1 + \frac{1}{2}\frac{d \ln F}{d \ln \dot{\gamma}_R}\right) \tag{31}$$

where $\dot{\gamma}_R = \Omega R/L$.

As in the case of the cone-and-plate system, F must be replaced by F′, where F′ includes a correction for inertial effects. Inertial stresses are far more significant in the parallel-plate system than in the cone-and-plate, at comparable shear rates. Because the shear rate varies spatially in the parallel-plate system, the inertial correction cannot be obtained exactly from experiments with a Newtonian fluid; however, the Newtonian stress estimate will considerably reduce the error due to inertial effects. Any residual anomalous behavior can be detected by plotting τ_{22} against the plate separation L, at constant shear rate. If the curve is not flat, it is suggested that the τ_{22} curve be extrapolated to zero plate separation, and that value of τ_{22} used as a corrected stress.

V. CONCENTRIC CYLINDER (COUETTE) SYSTEM

This system may also be used to measure normal stresses. The primary measurement is the difference in radial stresses exerted at the two cylindrical surfaces. As illustrated in Figure 4, this measurement is usually made by manometers.

If 1, 2, and 3 denote, respectively, the angular, radial, and axial directions, then the radial dynamic equation may be written as

$$-\rho \frac{v_1^2}{r} = -\frac{\partial P}{\partial r} + \frac{\partial \tau_{22}}{\partial r} + \frac{\tau_{22} - \tau_{11}}{r} \tag{32}$$

subject to the usual assumption of a laminar simple shear flow. If the inertial term is neglected, Equation 32 may be integrated between the inner and outer radii to give

$$\Delta(p - \tau_{22}) = -\Delta T_{22} = \int_{SR}^{R} \frac{\tau_{22} - \tau_{11}}{r} \, dr \tag{33}$$

Since $r^2 \tau =$ constant, τ may be substituted for the variable of integration, and the resulting equation may be differentiated with respect to the shear rate at the inner wall $\dot{\gamma}_i$. This leads to (after differentiation and use of the Leibniz rule):

$$\frac{d\Delta T_{22}}{d\dot{\gamma}_i} = \frac{1}{2} \left[\left(\frac{\tau_{22} - \tau_{11}}{\tau} \right) \frac{d\tau_0}{d\dot{\gamma}_i} - \left(\frac{\tau_{22} - \tau_{11}}{\tau} \right)_i \frac{d\tau_i}{d\dot{\gamma}_i} \right] \tag{34}$$

where quantities subscripted 0 and i are to be evaluated at the conditions of the outer and inner walls, respectively.

For a fixed geometry, since $\tau_0 = s^2 \tau_i$, it follows that

$$\frac{d\tau_0}{d\dot{\gamma}_i} = s^2 \frac{d\tau_i}{d\dot{\gamma}_i} \quad \text{or} \quad \frac{d \ln \tau_0}{d \ln \dot{\gamma}_i} = \frac{d \ln \tau_i}{d \ln \dot{\gamma}_i} = n \tag{35}$$

where n, as defined here, is the slope of the logarithmic plot of τ vs. $\dot{\gamma}$.

Equation 34 can be rearranged to the form

$$\frac{2}{n} \frac{d\Delta T_{22}}{d \ln \dot{\gamma}_i} = (\tau_{22} - \tau_{11})_0 - (\tau_{22} - \tau_{11})_i \tag{36}$$

The left side of this equation may be measured as a function of shear rate $\dot{\gamma}_i$, with $\dot{\gamma}_i$ calculated from Equation 33. As s approaches unity, the shear rates at the inner and outer walls approach each other, and the difference on the right side of Equation 36 becomes subject to considerable error.

A useful approximation is possible if s is not close to unity. Let $\tau_0 = s^2 \tau_i$, and suppose the viscosity behavior is approximated by the power law model

$$\dot{\gamma}_0 = s^{2/n} \dot{\gamma}_i \tag{37}$$

As a further approximation

$$(\tau_{22} - \tau_{11}) = A\dot{\gamma}^{2n} \tag{38}$$

Also, $(\tau_{22} - \tau_{11})_0 = s^4(\tau_{22} - \tau_{11})_i$ follows as an approximation. If $s = 1/2$, the normal stress difference at the outer wall is only $1/16$ of the value at the inner wall. Hence, for $s \langle 1/2$, a good approximation would be

$$\frac{2}{n}\frac{d\Delta T_{22}}{d\ln\dot{\gamma}_i} = -(\tau_{22} - \tau_{11})_i \tag{39}$$

As in the parallel-plate system, inertial stresses become important at high shear rates, and limit to some extent the range of shear rates that may be covered. Inertial stresses based on studies with Newtonian fluids may be used to correct the data, but only approximately, since the shear rate varies across the annular gap significantly when s is not near unity. If the power law model is assumed it can be seen that the ratio of apparent viscosities $(\tau/\dot{\gamma})$ at the inner and outer cylinders is

$$\eta/\eta_0 = s^{2(1/n - 1)} \tag{40}$$

If $s = 1/2$ and $n = 1/2$, then $\eta_i/\eta_0 = 1/4$, and the actual non-Newtonian inertial stresses might be poorly approximated by any Newtonian estimate. With this precaution in mind, it may be noted that a first-order estimate of the inertial effect on ΔT_{22} is given by

$$\mathscr{F} = \frac{\rho s^2 R^2}{8}(1 - s^4 + 4s^2 \ln s)\dot{\gamma}_i^2 \tag{41}$$

where $\dot{\gamma}_i$ is the Newtonian shear rate at the inner cylinder. From this expression it is possible to obtain an estimate for the inertial stress difference in a power law fluid,

$$\mathscr{F} = \frac{\rho s^{2/n} R^2 n^2}{8}\left[s^{2/n}(1 - s^2) - \frac{2s^{2/n}(1 - s^{2(1 - 1/n)})}{1 - 1/n} + \frac{s^{2/n}(1 - s^{2(1 - 1/n)})}{1 - 2/n}\right]\dot{\gamma}_{in}^2 \tag{42}$$

where $\dot{\gamma}_{in}$ is the power law shear rate.

If $\mathscr{F}$ is estimated, the foregoing equations are unchanged so long as ΔT_{22} is replaced by $\Delta T_{22} + \mathscr{F}$. If inertial stresses are effectively removed by this method, then data for $\tau_{22} - \tau_{11}$ should fall on a single curve, when plotted against shear rate, with no dependence upon s or R.

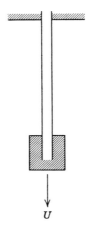

FIGURE 5. Definition sketch for a simple elongational flow.

VI. ELONGATIONAL FLOWS

Simple elongation can be defined as a flow with which another material property, the elongational viscosity, can be measured. One method of achieving a simple elongation is illustrated in Figure 5. The upper end of a cylindrical sample is clamped and stationary. The lower end is clamped and moves with the velocity U. If end effects in the region of the clamps can be ignored then the deformation will be uniform, in the sense that the shrinking of the radius will not depend upon axial position, and the axial velocity will be uniform across the radius. Hence, v_z is not a function of r, nor is v_r a function of z.

The axial velocity and radial velocity must satisfy the continuity equation:

$$\frac{\partial v_z}{\partial z} + \frac{1}{r} \frac{\partial}{\partial r} (r v_z) = 0 \tag{43}$$

In order to satisfy continuity, then it must be true that

$$\frac{\partial v_z}{\partial z} = \text{Constant} = -\frac{1}{r} \frac{\partial}{\partial r} (r v_r) \tag{44}$$

A solution which satisfies the boundary conditions $v_z = 0$ at $z = 0$, $v_z = U$ at $z = L(t)$, and $v_r = 0$ at $r = 0$, is

$$v_z = Uz/L \quad , \quad v_r = -Ur/2L \tag{45}$$

The length of the sample increases from its initial value L_o according to

$$L(t) = L_O + \int_0^t U(t)dt \tag{46}$$

Because the volume of the sample must remain constant, it follows that the area of the cross-section decreases with time, and is given by

$$A(t)/A_O = L_O/L(t) = 1/\alpha_1 \tag{47}$$

where α_1 is the "principal extension ratio".
The strain rate is

$$\dot{\epsilon} = \frac{\partial vz}{\partial z} = \frac{U}{L} = \left(\frac{1}{L}\right)\frac{dL}{dt} = \left(\frac{1}{\alpha_1}\right)\frac{d\alpha_1}{dt} \tag{48}$$

In the usual elongation experiment the velocity U is held constant, for example, $U = U_O$. Then ϵ becomes

$$\dot{\epsilon} = U_O/L = U_O/\alpha_1 L_O \tag{49}$$

and it is clear that the strain rate continually decreases as elongation proceeds. Because material functions may depend on $\dot{\epsilon}$, one must be careful in interpreting results from such an experiment.

A quantitative example of behavior under constant speed extension can be obtained by examining a simple linear viscoelastic model of the form

$$\tau + \lambda \frac{d\tau}{dt} = \eta_O\Delta \tag{50}$$

In the limit of vanishing relaxation time λ, one obtains the Newtonian fluid. For the axial direction, Δ_{11} is $2\dot{\epsilon}$, and hence is given by $\Delta_{11} = 2U_O/\alpha_1 L_O$. Since $L = L_O + U_O t$, it follows that $\alpha_1 = 1 + U_O t/L_O$. If the time variable is replaced by α_1 in the constitutive equation above and dimensionless variables are introduced, the resulting differential equation may be written as

$$\tau^* = \Lambda \frac{d\tau^*}{d\alpha_1} = \frac{1}{\alpha_1} \tag{51}$$

where

$$\tau^* = \tau_{11}L_O/2\eta_O U_O \quad \text{and} \quad \Lambda = \lambda U_O/L_O$$

The initial condition is simply $\tau^* = 0$ at $\alpha_1 = 1$. The solution for this model may be represented as τ^* vs. α_1 with Λ as a parameter. Figure 6 shows a

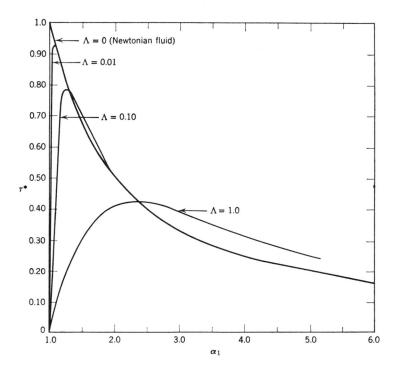

FIGURE 6. Comparison of Newtonian and viscoelastic elongational flows, at constant speed, plotted as reduced stress vs. principal extension ratio.

family of solutions. In the limit of vanishing Λ, the Newtonian solution, $\tau^* = 1/\alpha_1$, is approached.

A constant rate of strain experiment may be achieved by an elongation program given by

$$U = \beta L_o e^{\beta t} \tag{52}$$

where β is some constant. For this choice of U it is found that L is

$$L = L_o e^{\beta t} \tag{53}$$

and

$$\dot{\epsilon} = U/L = \beta \tag{54}$$

Hence, by suitable choice of β, the desired value of strain may be achieved.

VII. DYNAMIC MEASUREMENTS

The flows discussed up to this point have been those maintained under steady-state (time-independent) conditions. It is possible to carry out tests under dynamic conditions in which an external variable, such as the pressure drop across a capillary or the angular velocity of one cylinder of a Couette instrument, undergoes some programmed time variation. One then measures some features of the transient response of the fluid and infers therefrom the properties of the fluid under examination.

Two flows that particularly lend themselves to dynamic measurements are the oscillatory flow and the relaxational flow. In the former experiment the system has imposed upon it sinusoidal external conditions of known frequency and amplitude. The frequency, amplitude, and phase of the response of the fluid are measured. In a relaxational flow external conditions, such as force or strain, undergo a rapid change from one steady state to a second steady state. The response of the fluid as it approaches a new equilibrium state is then measured.

In both types of flows, by appropriate measurements, it is possible in principle to determine the viscosity and the normal stress coefficients for the fluid. If the experiments are properly designed and interpreted these coefficients should be identical to those measured in some steady-state configuration. It has been more common, however, to define new coefficients that are peculiar to the particular type of dynamic test undertaken. It is important, then, to investigate the possibility that these new coefficients might be related to the viscosity and normal stress coefficients, or otherwise to establish these coefficients as independent properties of the material.

Dynamic experiments are of interest for a number of reasons. Historically, because polymeric fluids are recognized to exhibit time effects such as stress relaxation, it has been natural to examine such effects directly through stress relaxation experiments. Furthermore, a very severe test of the applicability of a proposed constitutive equation is its ability to predict both steady-state and dynamic behavior. Such tests are often useful in differentiating among models which make similar predictions for steady-state flows, but differ significantly in predicted dynamic response. Finally, it is often possible to achieve a time scale in a dynamic experiment which is not easily accessible to a steady-state experiment. Hence, dynamic experiments are useful for extending the range of time scale over which a particular material is examined.

First, consider the problem of oscillatory flow. Consider a fluid filling the region between long concentric cylinders. Suppose the outer cylinder oscillates with frequency ω (s^{-1}) and amplitude ϑ_0. The angular velocity ω, the shear rate $\dot{\gamma}$, and the shear stress τ will exhibit sinusoidal behavior of frequency $\widetilde{\omega}$. In general these quantities will not be in phase with the input oscillation. If the inner cylinder is free to respond, for example, its angular velocity might be

$$\omega_i = \Omega_i \cos(\widetilde{\omega}t + \delta) \tag{55}$$

while the input oscillation would be

$$\omega_O = \Omega_O \cos\widetilde{\omega}t \tag{56}$$

A convenient representation of such oscillating quantities is in terms of complex exponentials. Equation 55 may be written

$$\omega_i = \text{Re}\{\Omega_i\, e^{i\,\widetilde{\omega}t}\} \tag{57}$$

where $\Omega_i = \Omega_{iR} + i\,\Omega_{iI}$ is a complex amplitude and Re { } means "the real part of { }". In terms of these quantities it may be shown that

$$\Omega_i = \sqrt{\Omega_{iR}^2 + \Omega_{iI}^2} \tag{58}$$

and

$$\delta = \tan^{-1}(\Omega_{iI}/\Omega_{iR}) \tag{59}$$

Similarly, the shear stress and shear rate are

$$\tau = \text{Re}\{Te^{i\,\widetilde{\omega}t}\} \tag{60}$$

and

$$\dot{\gamma} = \text{Re}\{\dot{\Gamma}e^{i\,\widetilde{\omega}t}\} \tag{61}$$

By analogy with earlier definitions one might define a viscosity coefficient as $\eta = \tau/\dot{\gamma}$. Because τ and $\dot{\gamma}$ are not necessarily in phase with one another such a definition leads to a time-dependent viscosity which then does not really represent a material property. Instead, it is common to define a complex viscosity coefficient as

$$\eta^* = T/\dot{\Gamma} = \eta' - i\eta'' \tag{62}$$

The parameter η' is the "dynamic viscosity" and is a function of frequency $\widetilde{\omega}$ in essentially the same way that the steady shear viscosity η is a function of shear rate. The parameter η'' is found to be a measure of elastic response of the material, and can be shown to be related to the steady shear normal stress coefficient Ψ_{12}.

A common method of operation is to have the bob suspended from a thin torsion wire. The outer cylinder, or cup, oscillates and, after initial transients damp out, a sinusoidal motion is induced in the annular body of fluid concentric with the cup and bob. A sinusoidal torque is exerted upon the bob which is opposed by the elasticity of the suspending wire. At equilibrium the bob oscillates with an amplitude and phase which depend upon the fluid properties, the amplitude and frequency of the cup motion, the moment of inertia of the cup, and the torsion constant of the wire.

The cup motion may be arbitrarily taken to be strictly cosinusoidal, so that ϑ_0 (angular deflection from equilibrium) is strictly a real quantity. The bob response is, in general, out of phase with the cup and so ϑ_i is a complex quantity: $\vartheta_i = \vartheta_R + i\,\vartheta_I$. The primary measurements consist of the amplitude ratio $q = |\vartheta_i|/\vartheta_0$ and the phase difference $\delta = \tan^{-1}(\vartheta_I/\vartheta_R)$. From these two quantities, at any particular frequency $\widetilde{\omega}$, the coefficients η' and η'' may be determined. By performing experiments over a range of frequencies, the frequency dependence of η' and η'' may be determined. This analysis leads to the following

$$\eta' = -\frac{Gq\,\sin\delta}{1 + q^2 - 2q\,\cos\delta} \tag{63}$$

$$\eta'' = \frac{G(q\,\cos\delta - 1)}{1 + q^2 - 2q\,\cos\delta} \tag{64}$$

where

$$G = \frac{I\widetilde{\omega}^2 - k}{4\pi L\widetilde{\omega}}\,\frac{R_O^2 - R^2}{R_O^2\,R^2} + \frac{\widetilde{\omega}\rho}{8}\,\frac{(R_O^2 - R^2)^2}{R_O^2}$$

It is also possible to perform an experiment in which the cup is fixed and the upper end of the torsion wire is twisted periodically through an angle $\psi = \psi_0 \cos \widetilde{\omega}t$. The bob then performs periodic oscillations of frequency but of different amplitude and phase, so that $\vartheta_i/\psi_0 = qe^{-i\delta}$. Again, measurement of the amplitude ratio q and the phase angle δ permits calculation of η' and η''. For this case

$$\eta' = \frac{R_O^2 - R^2}{4\pi\,LR_O^2\,R^2}\left[I\widetilde{\omega}^2 - k + \frac{k\,\cos\delta}{q}\right] \tag{65}$$

$$\eta'' = \frac{R_O^2 - R^2}{4\pi\,LR_O^2\,R^2}\,\frac{k\,\sin\delta}{q} \tag{66}$$

subject to the approximation $\rho\widetilde{\omega}R_O^2/|\eta^*| \ll 1$.

It is common practice among those who investigate solid-like materials to define material coefficients based upon the shear strain γ rather than the

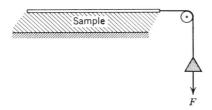

FIGURE 7. Definition sketch for a shear creep experiment at constant shear stress.

shear rate $\dot{\gamma}$. In a sinusoidal experiment γ would be written as

$$\gamma = \text{Re}\{\Gamma e^{i\,\widetilde{\omega}t}\} \tag{67}$$

However, because $\dot{\gamma} = d\gamma/dt$, γ could also be written in the form

$$\gamma = \text{Re}\{(\dot{\Gamma}/i\widetilde{\omega})e^{i\,\widetilde{\omega}t}\} \tag{68}$$

A "complex shear modulus" is defined as $G^* = T/\Gamma$. Since $\dot{\Gamma} = i\,\widetilde{\omega}\,\Gamma$ it follows that $G^* = i\,\widetilde{\omega}\,\eta^*$. G^* is usually written as $G' + iG''$, where G' is called the "storage modulus" or "dynamic rigidity", and G'' is called the "loss modulus". As noted above, $G' = \widetilde{\omega}\,\eta''$. It can be seen also that $G'' = \widetilde{\omega}\,\eta'$.

A "complex compliance" J^* may also be defined as $J^* = J' - iJ'' = \Gamma/T$. By simple manipulation it can be shown that $J' = G''/(G'^2 + G''^2)$ and $J'' = G''/(G'^2 + G''^2)$; J' and J'' are called the storage and loss compliances, respectively.

VIII. CREEP AND RELAXATION

It is important to note that a given material may show different classes of mechanical response under conditions involving different time scales. Consider, for example, a sheet of polymer suddenly subjected to a constant shear stress τ by the action of a force pulling on one plate bonded to the polymer; as in Figure 7, limiting responses are possible. If the material is a true elastic solid, then the shear strain γ immediately attains and remains at a constant value as long as the stress is maintained. If the stress is removed the strain relaxes instantaneously and the solid returns to its initial unstressed state. If the material, however, is a fluid, then the shear strain continuously increases with time as long as the stress is maintained. If the stress is suddenly removed the strain does not relax, but instead remains constant. An irreversible deformation, flow, has occurred. Figure 8 illustrates these responses.

In most polymeric materials neither of these simple responses occurs. Instead, a combination of elastic and viscous behavior gives rise to the strain

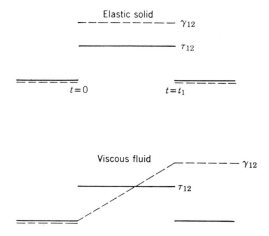

FIGURE 8. Hypothetical shear strain response to a constant shear stress imposed for a finite time. The perfectly elastic solid shows instantaneous and complete recovery of strain upon release of stress. The purely viscous fluid shows no recovery of strain.

history indicated in Figure 9. One associates such behavior with viscoelastic materials.

If the material is observed over a very short time scale compared to the time λ of Figure 8 it is likely that the observer would conclude that the material is an elastic solid. In contrast, if the time λ is very short compared to the response time of suitable measuring instruments, one is likely to conclude that the material under examination is a fluid. For many polymeric materials, at least in certain ranges of temperature and molecular weight, the complete response curve is accessible to measurement, and so viscoelastic parameters may be measured.

The response curve of Figure 8 is often described by a function called the "creep compliance", defined as

$$J(t) = \gamma(t)/\tau \qquad (69)$$

and is typically expressed as the sum of three parts:

$$J(t) = J_o + J\psi(t) + t/\eta \qquad (70)$$

J_o is the "ordinary elastic compliance" and represents the initial elastic response to stress. The term t/η represents the long time flow response and $\psi(t)$ is called the "retarded elasticity function", and is defined in such a way that it approaches unity as time increases indefinitely. J is referred to as the steady-state shear compliance, and like J_o and η, is a material constant for a given shear stress. The retarded elasticity function, while not a material

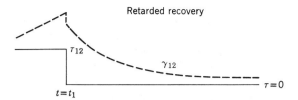

FIGURE 9. Hypothetical strain recovery upon release of a shear stress for a viscoelastic fluid.

constant, is nevertheless a characteristic property of a material. An experiment giving rise to a response such as is shown in Figure 8, is referred to as a "creep experiment".

If the material is in elastic equilibrium by virtue of some previous creep behavior followed by steady flow, and if the stress causing this flow is suddenly removed, then retarded elastic recovery occurs, as shown in Figure 9. The strain relaxation history is then the reverse of the creep curve.

Defining $\gamma(0)$ as the strain at the instant of stress removal

$$\gamma(0) - \gamma(t) = \tau[J_o + J\psi(t)] = \tau[J(t) - t/\eta] \tag{71}$$

where τ represents the constant stress that led to $\gamma(0)$. Note that in strain recovery there is no flow mechanism because the external stress no longer acts on the material. An obvious test of consistency of any creep experiment is the prediction of the recovery curve from measurements, under creep, of J_o, J, and $\psi(t)$.

The viscosity of the material can be obtained from the long time portion of the creep curve. Because it is practically impossible to measure viscosity of such rubber-like materials by simple techniques such as capillary flow or cone-and-plate shearing, which must be restricted to relatively low viscosities, the creep experiment provides an important extension of viscometry methods to the region of extreme viscosities. Because the time scale of the creep experiment is so low, these viscosities are "zero-shear" viscosities and hence the polymer is essentially Newtonian in its flow behavior under these conditions.

Another instrument used to measure viscoelastic properties is the torsional pendulum. A circular disk hung from a torsion wire is in contact with the material to be studied. The disk is displaced through a small angle about its axis, the torsion in the wire tends to restore the disk to its equilibrium position, but the inertia of the disk causes it to overshoot the equilibrium position. Thus, the disk performs torsional oscillations of fixed frequency and damped amplitude. The frequency and damping factor of the oscillations are measured, and may be related to instrument constants and properties of the material.

The creep experiments and their corresponding dynamic coefficients are related to shear behavior. Elastic materials are commonly studied in tension as well as in shear, and it is common to study viscoelastic materials in tensile experiments. Thus, one may study elongation (creep) under constant tensile load or tensile stress relaxation at constant tensile strain.

A simple relaxation experiment involves the sudden stretching of a strip of material to a fixed elongation. The tendency for creep is counteracted by continually reducing the force maintaining the elongation in such a way that the initial elongation is not altered. From the resulting data of stress vs. time it is possible to calculate a relaxation modulus.

As in the case of shear creep, tensile creep experiments may be used to measure the viscosity of semisolid materials. Viscosities as high as 1015 P can be measured. Tensile creep is similar to the simple elongation experiment. The difference lies in the fact that in elongational flows the strain is programmed and the measured response is the stress (force), while in the creep experiment the force is usually fixed and the measured response is the strain.

Chapter 5

RHEOMETRIC TECHNIQUES FOR POLYMER MELTS

I. PRACTICAL CONSIDERATIONS OF RHEOLOGY

The response of thermoplastics during processing can be categorized by three classes of property:

1. Deformation processes, which are necessary to form the product
2. Heat and heat transfer, which are necessary to achieve the plastic state
3. Chemical and physical change, which may be deliberately induced or adventitious

The subject of rheology is concerned with the study of deformation and flow: of all the responses of materials it is the one that is most readily observed. We have all squeezed toothpaste, kneaded bread dough, or tried to wipe glue from our fingers. As noted in earlier chapters, rheology is one way of describing those sensations.

The fundamentals of rheology are drawn from mechanics and provide the support for this study, which is mainly concerned with how materials systems differ from the ideals of classical mechanics. One approach to rheology is to broaden the field of classical mechanics by defining more generalized materials, the properties of which are derived logically from an equation of state. The advantage of this approach is that it offers full predictive capacity in any deformation history once the equation of state is described by appropriate experiments. At the other end of the spectrum lies the view that the interaction between complex materials and complex histories provides a series of unique situations that can only be studied in their own environment. An intermediate course is to evaluate the response of a material system in controlled experiments which are qualitatively similar to elements of its processing history and to discover from such experiments what the properties of the material appear to be.

Three material states are relevant to polymer processing: granular, the form in which materials are fed to the process; melt, the form in which they are usually shaped; and solid, the form of the final product, but also one in which some shaping may take place. This chapter deals with the rheology of melts, where most of the deformation occurs. However, in illustrating material property data some discussion on granular and solid response are included where these properties are relevant to processing.

Rheology is concerned with the relationship between stress, strain and time.

In reviewing some of the principles presented in earlier chapters, there are three simple deformations:

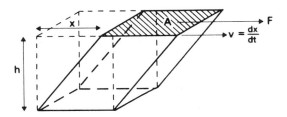

FIGURE 1.   Simple shear: area A and distance h remain constant during deformation.

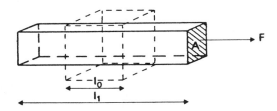

FIGURE 2.   Simple extension: cross-sectional area A and sample length l both vary during deformation.

1. In simple shear the stress is applied tangentially (Figure 1):

$$\text{Stress (N/m}^2)\qquad \sigma_s = F/A \tag{1}$$

$$\text{Strain (unity)}\qquad \gamma = x/h \tag{2}$$

$$\text{Rate of strain (s}^{-1})\quad \dot{\gamma} = \frac{1}{h}\frac{dx}{dt} = \frac{v}{h} \tag{3}$$

2. In simple extension the stress is applied normal to the surface of the material, as shown in Figure 2.

$$\text{Stress (N/m}^2)\qquad \sigma_E = F/A \tag{4}$$

$$\text{Strain (unity)}\qquad \epsilon = \int_{\ell_0}^{\ell_1} \frac{dl}{\ell} \tag{5}$$

$$\text{Rate of strain (sec}^{-1})\quad \dot{\epsilon} = v/\ell \tag{6}$$

3. In bulk deformation the stress is applied normal to all faces. The stress is the applied pressure, P, and the strain is the change in volume per unit volume, $\delta V/V$.

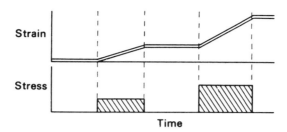

FIGURE 3. Newtonian behavior.

In the case of polymers, the response to a simple deformation process may be complex. Thus, in a simple shearing flow not only is there a shear stress, but also a pull along the lines of flow if usually described as the normal stress. The practical deformations of polymer processing are themselves usually complex flows compounded of shear, extension, and bulk deformations. One solution to this complexity is to introduce tensor notation which is the starting point of many rheological treatises. For the purposes of this discussion, it is sufficient to recognize and remember that those complexities exist; with that appreciation it is possible to seek simplifications in the practical response of real materials.

There are three types of response to an applied stress: viscous flow, elastic deformation, and rupture.

In viscous flow a material continues to deform as long as the stress is applied and the energy put in to maintain the flow is dissipated as heat. Viscosity is defined as the ratio of stress to rate of strain (units of Newton second per square meter).

If the only response of a material is to flow with a viscosity that is independent of stress level, the material behaves as a Newtonian liquid. At very low stress levels many polymer melts approach Newtonian behavior, which is illustrated in Figure 3.

If the material deforms instantly under stress and the deformation is spontaneously reversed when the stress is removed, the material possesses an elastic response. The modulus of such a material is the ratio of stress to recoverable deformation (units of Newtons per square meter). If all the deformation is reversible and the deformation is proportional to the applied stress the material is said to have a Hookean response (Figure 4). At low deformations rubbers approach Hookean behavior, and many polymer melts have a lower modulus than rubber.

Polymer melts may be characterized as having an elastico-viscous response to stress. The Maxwell model in which a Newtonian viscous dashpot is placed in series with a Hookean elastic spring (Figure 5a) is one concept; the response to stress is illustrated in Figure 5b. The rheology of such a material is characterized by two parameters: the viscosity of the dashpot and

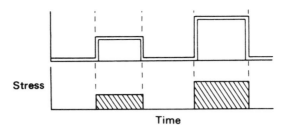

FIGURE 4.   Hookean behavior.

the modulus of the spring. The rheological response of polymer melts is more complicated. One approach to measuring that rheology is to assume that the material will respond as a Maxwell model and measure an apparent viscosity and an apparent modulus which depend on such factors as temperature, pressure, stress, geometry of deformation, and time. The properties which we record and use are thus the apparent Maxwell viscosity and apparent Maxwell modulus.

Polymer melts are commonly very brittle. Rupture plays an important role in determining the maximum rate at which a deformation process proceeds. In the practical response of polymer processing it is frequently the interactions between viscosity, elasticity, and rupture phenomena that determine the success of the operation.

In considering the bulk compressibility of materials the thermodynamic properties have already been acknowledged, and in viscous flow, where the work input is dissipated as heat, density and specific heat at constant pressure are immediately invoked. For a proper understanding of polymer processing we require a full appreciation of the pressure:volume:temperature relationships, or the material together with the heat transfer characteristics. It is important to note that polymers are excellent insulating materials. All those properties may be measured and handled by the techniques used for conventional materials.

The rate of heat exchange during processing is determined by thermal diffusion:

$$\text{Thermal diffusivity (m}^2\text{/s)} = \frac{\text{Conductivity}}{\text{Density} \times \text{Specific heat}} \quad (7)$$

which is readily applied using the concept of Fourier number:

$$\text{Fourier number} = \frac{\text{Thermal diffusivity} \times \text{Time}}{\text{(Section thickness)}^2} \quad (8)$$

(If heat is being exchanged from both sides of the section, the section thickness

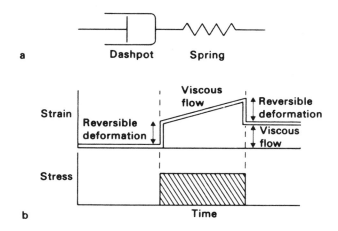

FIGURE 5. (a) Maxwell model; (b) its response to stress.

is halved.) Since heat transfer depends on the square of the thickness, a section doubled in thickness will take four times as long to reach thermal equilibrium. If the Fourier number is <0.1, the bulk of the material is very little affected by its environment; if it is >1.0, the whole of the material has come to near equilibrium with its surroundings. A value of about 0.3 indicates large temperature gradients. At first approximation the thermal diffusivity of polymer melts 10^{-7} m²/s. From this result it may be readily deduced that a section 2 mm thick, cooled from both sides, will approach thermal equilibrium in 10 s, while a 20-mm section will require about 15 min.

The concept of a polymer melt as a uniform amorphous continuum is idealized. One bulk polymer, polyvinyl chloride (PVC), is processed below its crystalline melting point and has been demonstrated to have a particulate flow mechanism. PVC is also usually plasticized and lubricated to achieve easier flow and may be further modified with "processing aids".

While PVC is near one extreme of the polymer spectrum, many other commercial polymers include some additional component that is essential either to their processing or to their service performance. Many commercial thermoplastic melts derive from semicrystalline resins and the thermal history of the process may allow some memory of that order to be retained. Finally, we have recently seen the introduction of "liquid crystal" polymers having a distinctive mesophase in the melt. The interaction of the morphology of a "melt" with its processing characteristics and the subsequent interaction between that processing history and the morphology — and so service performance of the end product is a subject of great interest in product development.

It is important to consider the chemical changes that may occur as a result of processing, which may range from significant modification of molecular weight of the whole polymer to specific surface conditioning.

The requirement for thermal stability depends on the type of process being operated, the scale of the process, and on the quality required in the finished product. For example, fiber spinning, polymers are deliberately degraded in order to reduce the molecular weight to obtain a very low viscosity polymer melt having a narrow molecular weight distribution. More often the requirement is for good thermal stability. The more sensitive processes are continuous extrusion processes and (especially) processes requiring a stretching of the melt — film blowing and blow molding. The least sensitive processes are fully constrained, intermittent flows such as injection molding.

The typical time scale for a processing operation is on the order 3 to 5 min from feeding granules to solidified product. This mean time masks a residence time distribution such that some 5% of the material will have a residence time on the order of 10 min, and a very small amount will have a very long residence time. The problem of residence time distribution is accentuated in processes such as sheet or film extrusion in which a pocket of degraded material may be disturbed from a dead space and locally lubricates the flow. Several commercial processes run at a high level of reclaim (blow molding is typically 50%) so that a simple residence time of 5 min becomes greatly extended when reclaim is taken into consideration. The problem of residence time distribution is also accentuated if the degradation mechanism is autocatalytic. Because of residence time distribution around protrusions, sometimes accentuated by catalysis from the metal parts, mechanical and appearance defects due to weld lines are common in easily degraded polymers.

The scale of a processing operation also alters the requirement for thermal stability. During extrusion from a 25-mm-diameter extruder, it is possible to maintain the melt temperature to within 5°C of the set barrel temperature and within 10°C of the melting point. In larger machines running fast (more typical of commercial practice), heat generation and low thermal conductivity result in a mean melt temperature usually at least 30°C above the melting point — local hot spots may be up to 20°C higher. Heat generation is a greater problem when working with polymers of high molecular weight. Running the extruder cold accentuates heat generation (and hence hot spots); for any material prone to degradation there is an optimum machine temperature.

II. RHEOLOGY AND POLYMER STRUCTURE

The rheology of a polymer strongly depends on the chemical and morphological structure of the material. If we consider the balance between service properties and ease of processing, a qualitative inverse relationship may be developed, as shown in Figure 6. That field shows three areas: along the line, the region of common experience; above the line, the target of a successful product; below the line, the disaster region. Movement toward the target may be achieved by altering the structure which may itself be modified by processing history. It is important to gain appreciation of the structural features that influence flow.

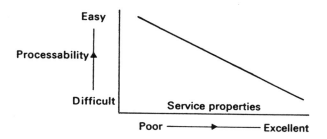

FIGURE 6. Relationship between service properties and ease of processing.

Part of this appreciation of how chemical structure influences chain flexibility, and thus rheology, may be obtained through molecular visualization of polymers, as shown in Figure 7. This means the molecule of polyethylene is easily recognized as highly flexible, with full freedom to rotate about each of the bonds. By comparison, polyparahydroxybenzoic acid has very little freedom of movement because of the rigidity of the aromatic links, and is intractable as a melt requiring powder metallurgy to shape.

In addition to the stiffening of the main chain, the molecule may be stiffened by adding side groups as in poly(methyl methacrylate), which greatly restricts the freedom of the molecule.

To determine the influence of chain stiffness and conformation on rheology quantitative description of molecular size is first needed. For a polymer of molecular weight M the length of the molecule, L, is obtained from

$$L = M(L_0/M_0) \tag{9}$$

where L_0 is the length and M_0 is the molecular weight of the repeat unit.

The space that a chain occupies, as distinct from its molecular volume, is obtained first from a consideration of the number of molecules, N, in 1 kg of polymer:

$$N = \frac{1000 \times 6.02 \times 10^{23}}{M} \quad , \quad \text{molecules/kg} \tag{10}$$

(6.02×10^{23} being Avogadro's number), in which the volume of one molecule, V, is given by

$$V = (\rho N)^{-1} \tag{11}$$

where ρ is the density.

The effective diameter of the chain, D,

$$D = \left[\frac{4}{\pi} (L_0/M_0)^{-1} (\rho \times 6.02 \times 10^{26})^{-1} \right]^{1/2} \tag{12}$$

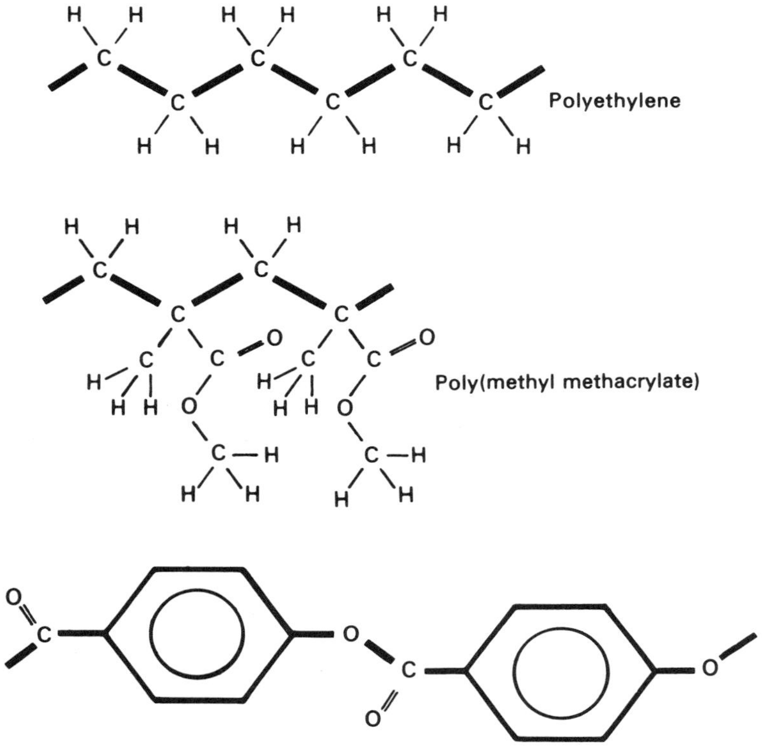

Polyethylene

Poly(methyl methacrylate)

Poly parahydroxy benzoic acid

FIGURE 7.   Examples of polymer chemical structure.

and $V = \pi D^2 L/4$.

Using the so-called polymer spaghetti model of molecular structure, we may anticipate a dependence on molecular length and diameter which, to a first approximation, may be normalized by considering a constant aspect ratio of length to diameter. Then, if A is the aspect ratio, we may deduce the molecular weight

$$M = AD(L_0/M_0)^{-1} \qquad (13)$$

Consider molecules of aspect ratio 1000 as typical of molecules having commercially useful properties. This aspect ratio is based on weight average molecular weight, M_w, for linear polymers of narrow molecular weight distribution ($M_w/M_n = 2$) (which is naturally obtained in the polymerization of many materials.

Taking a simplified view of the rheology, in the form of three parameters: η_O, the viscosity at a shear stress of 10^3 N/m^2 (Ns/m^2); G_0, the modulus at a shear stress of 10^3 N/m^2 (N/m^2); and $\sigma_{1/2}$, the shear stress at which η/η_O = 1/2. $\sigma_{1/2}$ is a description of non-Newtonian behavior, and from these terms we may derive the parameters

$$\tau_O = \eta_O/G_0 \quad ; \quad \gamma_C = \sigma_{1/2}/G_0 \tag{14}$$

The former is a characteristic of natural time, the latter is an elastic shear strain or orientation function.

Melt rheology depends greatly on temperature. As noted in earlier chapters it is thus necessary to make any comparison of rheologies on the basis of comparable temperatures. The choice of a common fixed temperature is not appropriate, because at any given temperature some polymers decompose while others are not molten. The choice of the temperature at which the melt is most commonly processed must be arbitrary; polyethylene, for example, is commonly processed at between 150° and 300°C. For this comparison we select a temperature 200°C above the glass transition temperature, T_g, a temperature that lies within the processing range of many commercial polymers and one at which the molecules may be considered to experience a common state of excitation.

The shear modulus, G_0, varies by more than an order of magnitude for different polymers. That difference is readily reconciled with what we know of the chemical structure. Taking polyethylene as a standard we may attribute the chain stiffness to the flexibility of the backbone. Other molecules [polypropylene, poly(4-methyl pentene-1), polystyrene] have the same backbone, but because of the bulky side groups, have a larger cross-sectional area. Because the stress can only be supported through the backbone we may anticipate an approximate inverse relationship between D^2 and G_0. We also see molecules, e.g., polyethylene terephthalate, which are much stiffer than polyethylene. Here, qualitative inspection of the chemical structure reveals the presence in the backbone of inflexible aromatic units that naturally stiffen the molecule.

The viscosity of melts of polymers of the same aspect ratio at 200°C above T_g likewise varies by an order of magnitude, but that variation is in direct correlation with the variation in modulus, so that the ratio η_O/G_0 is approximately constant. From this we can deduce that under these conditions viscosity depends only on the stiffness of the chain.

Further, the non-Newtonian character of the melt, as defined by the shear stress at which the viscosity has decreased to one half its low shear value, also varies directly with chain stiffness, implying a positive correlation between non-Newtonian flow and orientation.

The polymer that seems least consistent with the others, Nylon 66, is one that differs in molecular interaction by possessing strong hydrogen bonding. This almost certainly raises T_g and thus depresses viscosity at T_g + 200°C, and also has the effect of increasing non-Newtonian flow. A second polymer that falls outside the theory of the others is poly(methyl methacrylate), which appears to have a surprisingly high viscosity, possibly reflecting the stearic hindrance of the bulky side groups.

There are two common ways in which the viscosity of a polymer is reduced: plasticization and the incorporation of a more flexible comonomer unit in the backbone. Both courses reduce the glass transition temperature and hence the viscosity at a given processing temperature. The chain stiffness, and thus viscosity, is further reduced by increasing the flexibility of the backbone by the use of the comonomer, or by increasing the effective cross-sectional area of the molecule by the incorporation of plasticizer. Finally, the increased chain flexibility leads to an enhancement of non-Newtonian behavior and consequently a further reduction in viscosity under any given high shear condition.

Molecular weight is the single most important parameter in determining the viscosity of a polymer. A factor of two in molecular weight produces a tenfold change in the viscosity at a given shear stress in the range of molecular weight, M, of interest for thermoplastics. This relationship can be expressed as

$$\eta = kM^{3.5} \tag{15}$$

Equation 15 is true for a narrow range of molecular weights; a more appropriate relationship is

$$\eta = k_1 M + k_2 M^3 + k_3 M^5 + \ldots \tag{16}$$

In general, non-Newtonian flow is more pronounced in polymers of higher molecular weight. This feature may be qualitatively attributed to the argument that the greater the average number of entanglements per chain, the greater the probability of there being a region of intense entanglement which is broken down under lower average stress.

Molecular weight may be defined in many ways, of which weight average M_w and number average M_n are the most common:

$$\overline{M}_w = \Sigma M_i N_i^2 / \Sigma M_i N_i \tag{17a}$$

$$\overline{M}_n = \Sigma M_i N_i / \Sigma N_i \tag{17b}$$

The question as to which of these controls the viscosity is open. The answer can in part be gained by making physical blends of different molecular weights. For flexible polymers such as poly(4-methyl pentene-1), the longer chains appear to play the dominant role; for more rigid molecules such as poly(methyl methacrylate), it is the shorter chains that play the controlling role.

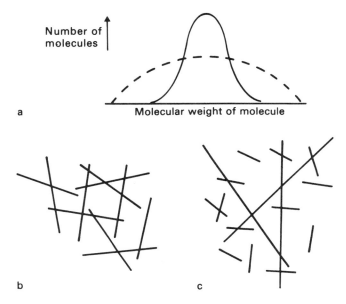

FIGURE 8. (a) Molecular weight distribution, broad MWD, narrow MWD; (b) narrow MWD, about 2 entanglements per chain; (c) broad MWD, long chain = 4 entanglements, short chain = 0 to 1 entanglements.

Polymers of lower molecular weight have lower viscosity and so are easier to shape. All other things being equal, polymers of higher molecular weight will have superior strength properties, thus the optimum polymer for any application requires a balance to be struck between processibility and performance. The expectation that service performance can always be enhanced by increasing molecular weight is confused by the fact that in polymer processing, all else is not equal, especially with respect to such features as the degree of orientation induced during processing. Orientation will tend to be higher when processing polymers of higher molecular weight and, while that orientation may enhance strength in the flow direction, this gain is commonly at the expense of a loss of properties in the transverse direction. There may thus be an optimum molecular weight to obtain the best possible service properties.

In addition to molecular weight, molecular weight distribution (MWD) is also an important contributor to rheology. MWD is usually represented by the polydispersity index as the ratio of weight to number average molecular weight.

When a sample has a narrow MWD, all the chains may be assumed to have a similar degree of entanglement so that they are uniformly stressed during flow. In a sample of broad molecular weight distribution, the longer chain molecules appear to form a protective network around their shorter brethren, as illustrated in Figure 8.

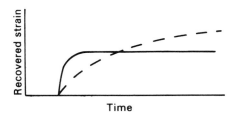

FIGURE 9.   Effect of MWD on strain recovery after steady flow-broad MWD, narrow MWD.

On the assumption that it is the entanglements that resist deformation, the long chains in the polymer of broad molecular weight distribution accept a disproportionately large amount of the total stress. This above average stress causes above average elastic response in those molecules, and this response is retarded by viscous resistance as the smaller molecules conform. Thus, in a sample of broad MWD there is usually more elastic response than in a sample of similar viscosity with narrow MWD, but that elastic response is delayed (Figure 9).

If it is stress that orients the molecules and thus reduces entanglements, then another consequence of the uneven stress distribution is that the entanglements in the longer chains start to disentangle at a lower average stress level. Thus, if a small high molecular weight tail is polymerized on top of a sample having low molecular weight the resulting flow curve shows two distinct transitions of non-Newtonian response: the first occurs at a low average stress when that average stress is largely taken by the few long molecules, placing them under high local stress; the second takes place at the same stress level as the unmodified low molecular weight polymer. The more general response for polymers of broad, rather than bimodal, MWD shows a more gradual transition.

Note that while the two polymers illustrated have approximately the same viscosity under melt flow index conditions, the viscosity at very low shear stress may be several orders of magnitude greater for the polymer of broad molecular weight distribution, while at high shear that polymer has significantly easier flow. In selecting materials for such applications as blow molding, high shear processibility and low stress form stability are commonly enhanced by broadening the MWD. However, these advantages are usually offset by poorer strength properties associated with the low molecular weight tail, and a loss of surface finish which may be associated with the small-scale heterogeneity of such materials. In processes such as injection molding the high elastic response and long relaxation times associated with polymers of wide molecular weight distribution lead, in addition, to frozen-in strain and the possibilities of warping, which must be set against the easier moldability.

Another important property is chain branching. As an example, if we consider low- and high-density polyethylene, these materials differ dramat-

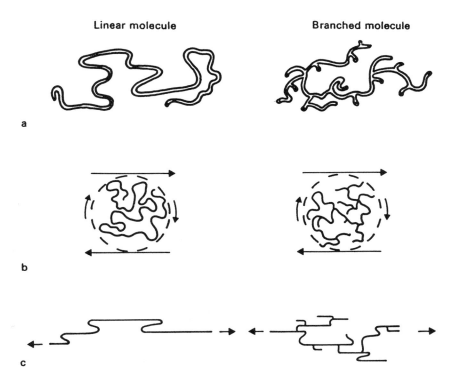

FIGURE 10.   (a) Branched and linear molecules; (b) shear flow; (c) stretching flow.

ically not because of any variation in monomer or MWD, but because of a difference in chain branching. High-density polyethylene is a linear molecule; low-density polyethylene made by high-pressure polymerization, is highly branched (refer to Figure 10a). The two polymers have similar shear flow behavior and similar elastic response but are dramatically different in stretching flows.

In shearing (rotational) flow is difficult to distinguish between branched and linear molecules (Figure 10b). If the branches are short the shear flow viscosity at a given molecular weight is depressed, but if the branches are long enough to entangle, then the viscosity is increased. However, when the molecules are highly ordered by a stretching flow the branch points act as hooks, increasing the resistance to flow (Figure 10c). The result is that while the linear polymer thins down under tension, the branched polymer tends to stiffen and finally rupture, as illustrated in Figure 11a.

Polymer melts that are superficially similar in other rheological tests may be 100-fold different in their resistance to high stress extensional flows. The tendency for the elongational viscosity of branched polymers to increase with stress means that local stress concentrations are less critical in establishing stretching flows (refer to Figure 11b).

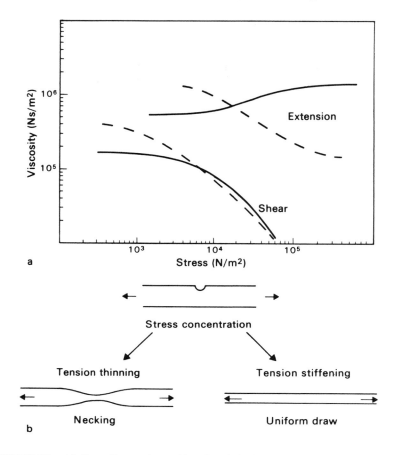

FIGURE 11. (a) Shear flow and stretching flow behavior of linear (——) and branched (- - -) polyethylene melt flow index 0.3 at 150°C; (b) effects of stress concentrations on stretching flows.

A further effect of branching, as evidenced by linear and branched polyethylene, is the dependence of viscosity on temperature. The viscosity of branched polyethylene is twice as sensitive to temperature as that of linear polyethylene. It is clear that it is not the short branches that are responsible for this change, since the temperature sensitivity of polypropylene, which could be considered a short-branched molecule with the same backbone, is the same as that of linear polyethylene.

Polymer melts that are superficially similar in other rheological tests may be 100-fold different in their resistance to high stress extensional flows. The tendency for the elongational viscosity of branched polymers to increase with stress means that local stress concentrations that are less critical in polymer melts exhibit an apparently liquid crystal mesophase, allowing them to be anisotropic at rest. Such polymers may be induced to develop a liquid crystalline state by the action of mechanical history under conditions of temperature and composition where that state does not spontaneously form.

Another situation involves block copolymers of the ABA type, in which surface tension forces drive the compatible ends to associate in star-shaped accretions. This structure is exploited in thermoplastic elastomers. The low stress rheology of such polymers is modified by mechanical history.

Melts from crystalline polymers tend to possess memory of that structure, recrystallizing in the same pattern after being subjected to temperatures several tens of degrees above their nominal melting points. Crystallization in such polymers is accentuated by high stress and pressure. In polypropylene, for example, crystallization may occur during processing at temperatures well above the nominal melting point, leading to dramatic increases in the resistance to flow.

III. RHEOMETRIC APPLICATIONS

Different categories of measurement may require different approaches to rheometry. Before embarking on any measurement it is necessary to decide on the goals of the measurement.

Subjecting a polymer to a sinusoidal shear history and measuring the stress response is established as a useful way of obtaining precise information about the rheology of the melt. If the material is entirely elastic the stress response is sinusoidal and in phase with the strain input, and has an amplitude dependent on the stiffness of the material. If the material is entirely viscous the stress response is exactly $\pi/2$ out-of-phase. No melt is entirely viscous or entirely elastic, so the response involves a phase shift (Figure 12). It is common for the response at low frequencies to approximate to viscous flow, while that at high frequency is usually far more elastic; thus the phase shift and the derived dynamic viscosity and modulus are frequency dependent. Generally those properties are not amplitude dependent because the maximum strain and strain rate in the measurement are sufficiently small that departures from linearity are rare.

This class of experiments provides precise data characterizing the rheology of the melt. There are, however, several ways of representing such data. The differences in interpretation rest on whether stress or strain is assumed to be the controlling mechanism. Most equipment actually operates under imposed strain, but the response would be identical if stress were the driving mechanism. Most instruments operate under the principle that strain is assumed to be the controlling variable, thus yielding the dynamic viscosity:

$$\eta' = \frac{\sigma^*}{\omega\gamma^*} \sin\delta \tag{18}$$

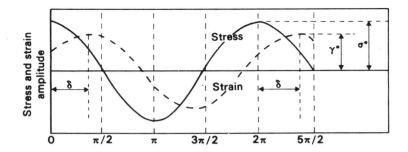

FIGURE 12.   Measurement of stress response, σ^* where γ^* is the strain amplitude and δ is the phase lag.

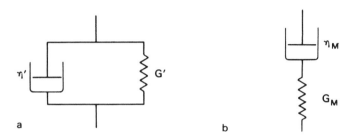

FIGURE 13.   (a) Voigt model; (b) Maxwell model.

and the modulus:

$$G' = \frac{\sigma^*}{\gamma^*} \cos\delta \qquad (19)$$

where σ^* and γ^* are the stress and strain amplitudes, δ is the phase lag, and ω is the angular velocity.

If the dynamic viscosity and modulus are independent of frequency we can deduce that the material is behaving similar to a Voigt element (Figure 13a). Because these parameters depend on frequency we may alternatively describe them as the apparent Voigt viscosity and modulus of the material. An alternative representation of the same data may be made using the stress as the controlling variable giving

$$\eta_M = \sigma^*/\omega \ \gamma^* \ \sin\delta \qquad (20a)$$

$$G_M = \sigma^*/\gamma^* \ \cos\delta \qquad (20b)$$

where, if η_M and G_M are independent of frequency, the material is responding like a Maxwell model (Figure 13b) and the properties η_M and G_M may be described as the apparent Maxwell parameters.

In selecting the most useful method of representation one may choose between that which is most commonly used —the apparent Voigt parameters — and that which most closely approximates to the behavior of the system that is being measured. The Voigt element is a solid in that it will not deform continuously under a sustained stress. The Maxwell element is a liquid, and, at sufficiently low frequency, most polymer melts approximate to a Maxwell model. Steady-flow rheological experiments on polymer melts implicitly use the Maxwell rather than the Voigt interpretation.

Most commercially available oscillatory shear rheometers operate in the range 10^{-3} to 10^3 rad/s. The lower limitation is one of time scale of measurement and the upper limitation is the complication arising from the inertial effects. Crystal excitation has been used to study the behavior of liquids at very high frequencies. Most dynamic measurements operate with strain amplitudes of the order 0.1.

The principal advantages of dynamic measurements for the study of the rheology of melts are:

• The techniques have been rigorously analyzed and may be used to obtain precise data in the linear region over a very wide range of frequency.
• They provide direct measurement of elastic as well as viscous properties, and for materials of low viscosity they may be the only practical way of comparing elastic response.
• Through the analogy between inverse frequency and time, dynamic methods yield information relevant to the very short time scale response of melts.

Disadvantages in using this type of measurement are:

• The uncertain correlation between the rheological response in the linear elastico-viscous region and the evident nonlinear response apparent in most polymer processing.
• The relatively high cost of the equipment.

In earlier chapters, rigorous treatment has been given to the more common types of rheometers. We will now discuss practical limitations of these instruments. The first of these is the cone-and-plate rheometer. There are two major variants, depending on whether the apparatus is designed to operate under constant stress or constant rate. Both are illustrated in Figure 14.

The constant-rate instrument, in which one face is rotated at a series of predetermined rates and the torque is measured on the opposite face, is preferred in most commercial instruments. If the instrument is being used as a viscometer only, this method of operation has an advantage over the constant stress variant in that when the system reaches steady flow, the torque output is a constant reading.

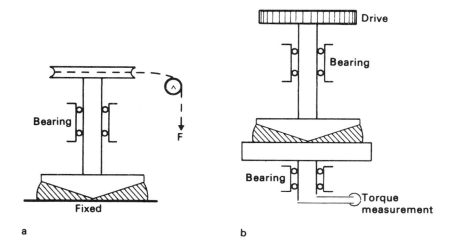

FIGURE 14.   Cone-and-plate rheometers. (a) Constant-stress type; (b) constant-rate type.

Cone-and-plate instruments may also be used to determine the elastic response of melts. Several measurements can be made, including stress and strain transients to and from steady flow, normal stresses, and strain recovery.

When an elastic liquid is subjected to simple shear flow, as well as the shear stress there is a pull along the lines of flow. In a cone-and-plate rheometer pull strangulates the flow, causing a pressure profile across the face of the instrument and a normal thrust forcing the faces apart. That thrust is most conveniently measured by instruments in which the strain rate is imposed, which accounts for the common preference for that class of instrument.

The measurement of the stress transients during start-up and cessation of steady shear flow offer opportunity for deducing the elastic response. Such transients can be of particular value for comparative measurements, but absolute measurements demand very accurate setting of the apparatus. Many observers report an effect usually described as the stress overshoot phenomenon when, on start-up of flow at a constant rate, the stresses go through a maximum before reaching its equilibrium condition (Figure 15). The phenomenon can be extremely reproducible and some workers deduce from it features of the viscoelastic response.

A particular advantage of the constant stress cone-and-plate is the ease with which the stress may be removed instantaneously, allowing the observer to record strain recovery as a function of time for different stress histories. The ratio of shear stress to recovered shear makes a simple comment on the elastic response which may be related to practical observations of frozen-in strain in moldings and postextrusion swelling, in that all represent a desire of the material to revert to an earlier state. The use of recoverable shear as the prime measure of elastic response has the additional advantage that while it is being measured, it is the only thing that is happening.

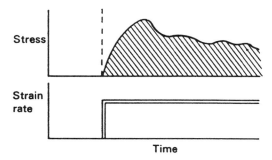

FIGURE 15. The stress overshoot phenomenon: stress passes through a maximum and subsequently decays.

Another advantage of the use of constant stress compared to an experiment at constant rate is that it requires a shorter time to reach equilibrium response. This difference in time scale is most clearly demonstrated for a Maxwell model in which, under constant stress, steady flow is attained instantaneously, while under constant strain rate the approach to equilibrium conditions is very slow. This is illustrated in Figure 16.

A final advantage of the constant stress method of operation is that most rheological phenomena, in particular the reaching of a defined recoverable strain and the onset of non-Newtonian flow, tend to occur at a near-constant shear stress rather than constant shear rate for a range of temperatures and molecular weights. Comparative testing of elastic phenomena is thus uncomplicated by viscosity. Further, comparisons of viscosity in shear thinning systems at constant shear stress are more discriminatory than comparisons at constant shear rate in non-Newtonian systems.

The advantages of cone-and-plate rheometry are:

- It gives a precisely defined flow at low shear stress.
- It can provide data on both elasticity and viscosity in a variety of experiments.
- As a measure of rheology under low stress it is especially sensitive to the structural characteristics of the melt, the structure of which may break down under higher stress.

The limitation of this class of rheometry for melts is associated with the very large ratio of surface area to volume of the instrument, with its attendant problems of slip that limit its usefulness to relatively low stress measurements. Even if it were possible to overcome the problems associated with the polymer-metal interface, the rate of heat generation would still preclude effective operation of the instrument at higher stress.

The second class of instruments to discuss are elongational devices. As described in earlier chapters, the principle involves the use of simple

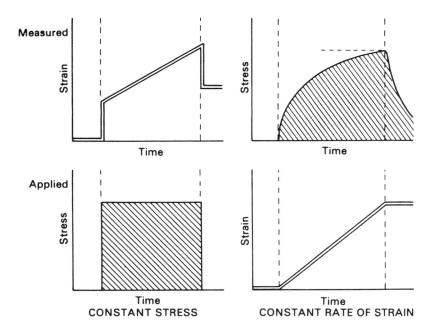

FIGURE 16. Deformation of Maxwell model: under constant stress steady flow is attained instantaneously.

instruments that have provided useful information on stiff melts. More sophisticated machines can supply high-quality data on a broader range of materials. Direct elongational flow measurements have a particular advantage over shear measurements in that the measurement does not involve an interface with the wall of the instrument, so that slip is not a problem. Such measurements are of direct practical importance as representative of the important drawing operations of practical processing and are of considerable theoretical significance as representing the class of rotational flows which are fundamentally different from shear flows.

The advantages of constant stress or constant rate as the controlling mode of deformation apply to extension as well as shear. The reduction of the time transient available when making measurements at constant stress is a special advantage in elongation flow when making measurements on melts whose viscosity decreases with increasing tensile stress. In such experiments the small residual unrecovered strain from a brief creep and recovery experiment (Figure 17) may be interpreted to give an apparent rate of viscous deformation in circumstances in which waiting for steady flow conditions to be achieved would allow the sample to neck.

The most important argument in favor of making direct measurements of elongational flow is that only by making such measurements can one satisfy oneself of the significant differences between shear and extensional flows and between different materials in such flows.

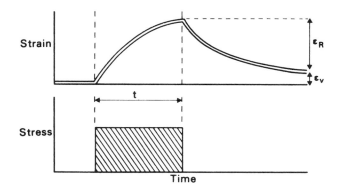

FIGURE 17. Elongational state rate.

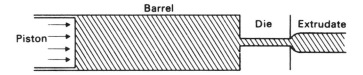

FIGURE 18. Capillary extrusion rheometer.

The subject of measuring apparent rheological properties via capillary flow was discussed in earlier chapters. Additional comments on practical limitations of this instrument follow.

By the selection of appropriate equipment it is possible to measure shear stresses in the range 10^3 to 10^6 N/m^2 and shear rates in the range 0.001 to 500,000 s^{-1}.

Two modes of operation have been used: controlled pressure by dead weight loading or gas pressure (requiring a measurement of the volumetric flow rate), and controlled volume displacement (requiring a measurement of pressure) (Figure 18).

What is required for the measurement is the relationship between flow rate and pressure drop through the die, and because pressure drops may be associated with the flow in the extruder barrel (and, more particularly, frictional losses between the piston and the barrel wall), it is highly desirable to make the measurement of pressure drop directly in the melt just above the die rather than relying on inferring pressure drop from the force required to drive the piston. The choice of barrel size is determined by the following considerations:

- The available charge
- The heat-up period to an equilibrium temperature (especially when the polymer is subject to degradation)
- Engineering tolerances

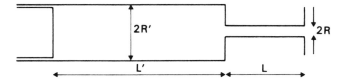

FIGURE 19.   The barrel-height effect, barrel radius R', length L'; die radius R, length L.

The larger the reservoir size, the more convenient it will be to engineer. The larger volume flow rates available from larger pots allow the same shear rate range to be covered with larger radius dies, which themselves will have better tolerances so that viscosity (proportional to R^4) may be more accurately determined. (In general, dies of radius 0.25 to 2 mm are found to be most convenient.) Unfortunately, larger pots require longer heating-up times and larger charges and it is this factor that limits the size appropriate to laboratory measurements.

Interpretation of capillary rheometry data using the Poiseuille equation yields an apparent rather than a true measure of viscosity at the operating temperature. A series of corrections is appropriate to derive the true viscosity, but only one of these, the ends (or Bagley) correction, is recommended.

It is further recommended that the long die used in such measurement should have a length-to-radius ratio of between 20 and 50, and preferably 32.

The principal sources of error when measuring high viscosity by capillary flow include:

- Reservoir and friction losses
- End pressure drop (Bagley correction)
- Nonparabolic velocity profile (Rabinowitsch correction)
- Slip at the die wall
- Influence of pressure on viscosity
- Influence of pressure on volume
- Influence of heat generation
- Influence of decompression on temperature
- Modification of the material due to work in the die

Corrections for the first two of these are recommended. The others, particularly those due to pressure and heat generation, may be larger but are to a considerable extent mutually canceling. Further, these other corrections are also a feature of most practical flows; thus, while we must certainly be aware of them, we may, in engineering calculations, ignore them and anticipate that their effect is already included in the data used.

The barrel-height effect derives from the fact that a capillary rheometer is actually two capillaries in series (Figure 19) and that during a test the effective length of the barrel (L') changes so that not only is there a correction

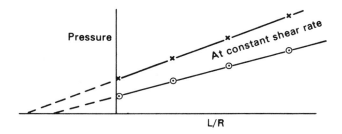

FIGURE 20.   Bagley correction.

for flow in the reservoir, but that correction is not constant. This effect, which can be particularly large with strongly pseudoplastic melts, is most conveniently avoided by inserting a pressure transducer just above the die. If this is not possible, measurements should be made within prescribed limits of piston height, when the measured end effect will include frictional and reservoir flow losses.

Piston friction may represent a source of error if the pressure measurement is deduced from load on the piston. Friction may be eliminated by gas pressure-driven extrusion, but the pressures available are themselves limited to about 2×10^7 N/m² (200 atm). A first correction for piston friction may be made by running without a die in place at all, but this takes no account of the fact that the friction is likely to be a function of the pressure inside the chamber.

Piston-height effects and friction are especially significant when making measurements of the flow through dies of very low length-to-radius ratio.

In the Bagley method, one finds that a plot of pressure drop vs. die length-to-radius ratio at a fixed wall shear rate gives a straight line with an intercept (Figure 20), such that the shear stress in capillary flow is more correctly $\tau = R/2(dP/dL)$. The pressure gradient may be ideally determined by using dies of several different lengths, but this requires extensive experimentation. In practice, the choice of two dies may be adequate. The greatest sensitivity is obtained by combining a long die (L/R = 32) with an orifice (L/R = 0) such that

$$\tau_S = (P_L - P_O)R/2L \qquad (21)$$

where P_L is the pressure drop through a long die and P_O is the orifice pressure drop. Some experimenters detect nonlinearities in the Bagley plot near L/R = 0, but the measured values of P_O are rarely found to be >20% from the extrapolated value at L/R = 0; larger differences usually indicate structural changes as a result of flow, or some other source of error is present. Since the pressure drop in the long die is commonly ten times that in the orifice die, the potential error in shear stress from such a nonlinearity does not exceed 2% and should not detract from the use of the orifice die for this correction.

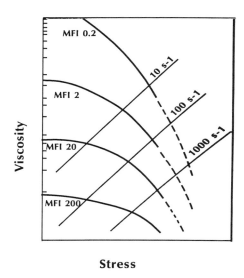

Stress

FIGURE 21.   Apparent viscosity vs. wall shear stress: low-density polyethylene at 170°C.

Second, the use of the maximum difference in L/R ratio for the two dies will greatly improve the reproducibility of the measurement of shear stress, since the cumulative error in pressure drop is reduced.

Finally, the measurement of orifice pressure drop, which may be represented as a function of corrected shear stress, is itself a useful rheological parameter.

For the measurement of apparent viscosity by capillary flow the ends correction should be made and the method of making that correction should be noted with the data. An appropriate method is the two-die method using one die with L/R = 32 and the other an orifice of the same diameter.

The data may conveniently be pressed as a graph of log apparent viscosity vs. log wall shear stress (Figure 21). Graphs of log apparent viscosity vs. shear log stress are recommended because such representations of flow curves, at different temperatures or pressures, for a particular polymer are usually superposable by a simple verticle shift. In addition, the flow curves of polymers of the same family but different molecular weights are also approximately superposable by a shift at constant stress, making it relatively easy to construct approximate additional data from a limited presentation. Further, values of viscosity or shear stress at fixed shear rates may be easily read from such graphs by reading along lines at 45°. Graphs of log apparent viscosity vs. log wall shear rate or of log wall shear stress vs. log wall shear rate may be used as alternative methods of presentation.

The ends correction or orifice pressure drop is a useful measurement in its own right. If plotted as entrance pressure drop against shear stress in fully

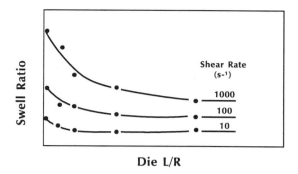

FIGURE 22. Effect of die length-to-radius ratio on post-extrusion swelling: polypropylene melt flow index 230/2 = 15 at 210°C.

developed flow the result is largely independent of melt temperature and molecular weight, but extremely sensitive to so-called structural variations such as MWD, chain branching, fillers, grafting, and particulate nature. The orifice pressure drop may also be used to obtain a qualitative assessment of the elongational flow rheology; the exit pressure drop can be used to develop a correlation between die exit pressure drop and postextrusion swelling.

The most obvious observable elastic effect during capillary extrusion is postextrusion swelling. Swelling is essentially a reversion toward an earlier state and as such is commonly interpreted as evidence of recoverable strain. Postextrusion swelling generally increases as flow rate increases and decreases toward a constant value as die length-to-radius ratio increases (Figure 22). To make a precise measurement of swell ratio, many correction factors must be taken into account. For practical purposes, it is the comparative measurement which is most important, and an adequate measurement may be obtained by cutting the extrudate flush with the die surface, extruding a short length, and then measuring the diameter of the solidified extrudate within 5 mm of the leading edge. The two measurements of swell ratio from an orifice die and from a die of length-to-radius ratio 32 approximate to the maximum and minimum values of swell ratio.

There are several accessories that can be adapted to a capillary rheometer to provide even more detailed information on the rheological properties of a melt. The first of these involves stretching flow, which is an obvious accessory to an extruder with an instrumented haul-off. Measurements of in-line tension, drawing profile, and rate can be of great value in comparing the stretchability of polymers. Simply clipping a weight on a sample and allowing it to draw until it solidifies is a simple method of comparing highly viscous materials in stretching flows. The main drawback of such measurements is the difficulty of interpreting the flows, with velocity gradients and complex history and temperature gradients, to yield fundamental data. The only parameters that are easily defined are the draw ratio and maximum stress to which the material

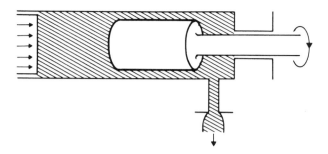

FIGURE 23.   Illustrates preshearing.

has been subjected (force/area in thinnest section). The measurement is thus most suitable for studying phenomena such as rupture in which maximum stress is the most important factor, or for comparative views of the extensibility of materials.

The melt from a capillary rheometer may also be extruded past a chamber where the material is subjected to a well-defined shear history, as illustrated in Figure 23. Such a section may be used for preparing melts of different shear history so that the influence of this important, though often neglected, parameter may be evaluated in separate experiments.

In addition to mechanical measurements of stress and strain, it is possible to make direct measurement of the streamlines and of stress birefringence during flow. The measurement of stress birefringence requires sophisticated equipment, experimentation, and interpretation, but the information yielded is comprehensive. The introduction of markers allows a qualitative judgment to be made of the streamline pattern.

The melt flow index test is a quality control test used to assess the fluidity of a melt under standard conditions. As a single point test, melt flow rate is extremely sensitive in distinguishing between members of the same poly-merization family. Under ideal conditions it can give reproducible values to within 3%, discriminating variations of 1% in molecular weight (on polymers whose molecular weight is on the order of 100,000). This sensitivity does not necessarily distinguish polymers that have been made by different routes in which the viscosity of polymers having the same melt flow rate may vary by more than an order of magnitude under other conditions.

Of all the members of the family of capillary flow systems, the melt flow indexer is the most widely used. It is also capable of much more versatile use than the measurement of a single value. Flow rate may be determined as a function of applied load so that a qualitative indication of pseudoplasticity is readily obtained by comparing the ratios of flow rate at two different loads. The extrudate is available for elasticity determination by means of the swelling ratio and may also be subjected to simple drawing experiments. The measurement of melt flow rate and swelling ratio before and after processing, especially at a low temperature, can be an especially useful indication of changes that occur during processing.

Chapter 6

TORQUE RHEOMETERS AND PROCESSABILITY TESTING

I. INTRODUCTION AND GENERAL INFORMATION

Torque rheometers with miniaturized internal mixers (MIM) are multipurpose instruments that are well suited for formulating multicomponent polymer systems, and studying flow behavior, thermal sensitivity, shear sensitivity, batch compounding, etc. The instrument is applicable to thermoplastics, rubber (compounding, cure, scorch tests), thermoset materials, and also to liquid materials.

The rheometer with a single screw extruder allows the measurement of rheological properties and extrusion processing characteristics to differentiate lot-to-lot variance of the materials. The rheometer also enables the process engineer to simulate a production line in the laboratory and to develop processing guidelines. Because all the measurements can be made by simulating production, it is much easier to solve problems that the engineer may face.

The torque rheometer with a twin screw extruder is considered a scaled-down continuous compounder. It allows the compounding engineer to develop polymer compounds and alloys. It also permits the formulation engineer to ensure his formulation is the optimum.

Because the instruments are very versatile, low viscosity (liquid) and high viscosity materials, high temperature engineering thermoplastics, rubber, and thermoset materials can be tested. Therefore, they are widely used in quality control and research and development areas in the polymer industries.

A torque rheometer is an instrument which is widely used in studying formulations, developing compounds, and characterizing the polymer flow behavior by measuring viscosity-related torque caused by the resistance of the material to the shearing action of the plasticating process.

Torque is the effectiveness of a force to produce rotation. It is defined as the product of the force and the perpendicular distance from its line of action to the instantaneous center of rotation.

A microprocessor-controlled torque rheometer is a very unique system, consisting of two basic units: electromechanical drive unit, and a microprocessor unit. Its advantages are:

1. The microprocessor-based torque rheometer can control two pieces of equipment simultaneously. An example of this would be a feeder, or postextrusion system. It allows the user to regulate feed rate or downstream equipment.
2. The system can drive equipment, control and monitor parameters, and has the additional feature of data acquisition capabilities.

3. The unit displays all parameters graphically or numerically on-line during testing.
4. The rheometer also has the capability to plot any parameter against another in the off-line mode.
5. One can make his own program using either on-line or off-line during testing.
6. The unit has a safety cut-off system when the torque or totalized torque exceeds maximum values.
7. All data can be printed on a printer on-line or off-line.
8. The rheometer will monitor and record several pressures and melt temperatures, as well as torque, totalized torque, and speed(s).
9. The rheometer is able to program several speeds and temperatures for studying shear sensitivity or temperature sensitivity.
10. The system allows one to magnify specific portions of the test data in order to better differentiate small differences between data.
11. Systems have automatic calibration capabilities for the torque load cell and pressure transducers in any range.
12. Every parameter measured is recorded for later reference.

Figure 1 shows a block diagram of a typical system.

II. MINIATURIZED INTERNAL MIXER (MIM)

Most polymeric products are not pure polymers, but mixtures of the basic polymer with a variety of additives, such as pigments, lubricants, stabilizers, antioxidants, flame retardants, antiblock agents, cross-linking agents, fillers, reinforcement agents, plasticizers, UV absorbants, and foaming agents. All these additives must be incorporated into the polymer prior to fabrication. Some of the additives take a significant portion of the mixture; others only minute amounts. Some are compatible; others are not.

Depending on the quality of resin and additives and homogenization of the mixtures, the quality of the final product will be varied. Therefore, developing a quality resin and additives when met with desired physical and mechanical properties of the product and quality control associated with them plays an important role in the plastic and rubber industries.

The mixer consists of a mixing chamber shaped like a figure eight, with a spiral lobed rotor in each chamber. A totally enclosed mixing chamber contains two fluted mixing rotors that revolve in opposite directions and at different speeds to achieve a shear action similar to a two-roll mill.

In the chamber, the rotors rotate in order to effect a shearing action on the material mostly by shearing the material repeatedly against the walls of the mixing chamber. The rotors have chevrons (helical projections), which perform additional mixing functions by churning the material and moving it back and forth through the mixing chamber. The mixture is fed to the mixing

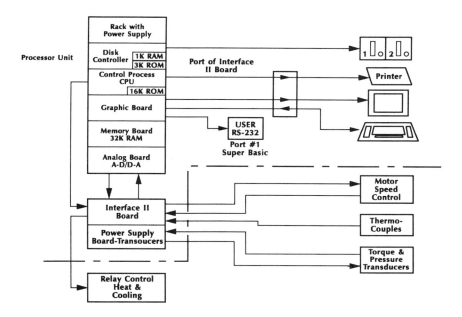

FIGURE 1. Schematic block diagram of a microprocessor-controlled torque rheometer.

chamber through a vertical chute with a ram. The lower face of the ram is part of the mixing chamber.

There is a small clearance between the rotors, which usually rotate at different speeds (e.g., gear ratio 3:2, 2:3, and 7:6 for different mixers) at the chamber wall. In these clearances, dispersive mixing takes place. The shape of the rotors and the motion of the ram during operation ensure that all particles undergo high intensive shearing flow in the gaps (clearances).

Normally, roller rotors are used for thermoplastics and thermosets, cam rotors for rubber and elastomers, and sigma rotors for liquid materials. Banbury rotors are used with a miniaturized Banbury mixer for rubber compounding and formulation. Figure 2 provides a schematic diagram of the MIM mixer.

Temperature is the most important factor in measuring rheological properties and/or flow behavior of the polymer mixes, because one can get different results with the same material by changing the temperature even by a few degrees.

The mixer consists of three sections, and each section is heated and controlled by its own heater and temperature controller. It is designed as such to maintain very accurate and uniform temperature profiles throughout the mixer.

Since mechanical dissipation heat is developed in the small gap between rotors and chamber, the heat conducts to the center bowl and raises the set

1. Back Section
2. Center Bowl
3. Front Plate
4. Rotor Shafts
5. Rotors
6. Air Cooling Channels
   Cast in Aluminum
7. Heaters Cast in Aluminum
8. Melt Thermocouple
9. Air Valve and
   Metering Plate
10. Air Exhaust
11. Bushings
12. Ram

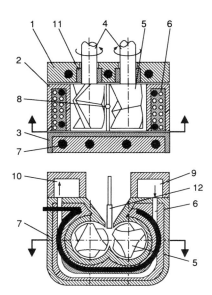

FIGURE 2.   Schematic diagram of MIM.

temperature. In this case, the heater at the center bowl is automatically shut off, and the cooling solenoid valve is automatically energized to circulate cold air through the center section; the heaters in the back and front sections remain energized to maintain their own set temperature through the test.

If the mixer was to consist of two sections with one heater and temperature controller only at the bowl, when the heat is shut off due to the conduction heat from the material, there is virtually no heating at the time; therefore, the back section becomes cold. This temperature control problem causes a non-uniform temperature profile throughout the mixer. Additionally, heating takes longer than that of the mixer with three heaters.

The loading chute is also designed for quick loading, and the chute can be removed from the mixer after the material is fed into the mixer. If the cold chute is not removed, the chute will draw the heat out of the mixer and the set temperature will be disturbed. This will not allow the measurement of flow behavior at the desired temperature. Additionally, if the chute cannot be removed after loading, the chute will be hot; it will cause problems in loading for the test to follow.

It is well known that plastic materials with two or more components being processed should be well mixed, so that the compounded material would provide the best physical properties desired for the final product.

There are distinctive types of mixing processes. The first is the spreading of particles over positions in space (distributive mixing) and the second is shearing and spreading of available energy of a system between the particles themselves (dispersive mixing). In other words, distributive mixing is used for any operation employed to increase the randomness of the spatial distribution of particles without reducing their sizes. This mixing depends on the flow and the total strain, which is the product of shear rate and residence time or time duration. Therefore, the more random the arrangement of the flow pattern, the higher the shear rate; the longer the residence time, the better the mixing.

A dispersive mixing process is similar to that of a simple mixing process, except that the nature and magnitude of forces required to rupture the particles to an ultimate size must be considered[1].

A more important function of the intensive mixing is incorporation of the pigments, fillers, and other minor components into the matrix polymer. This mixing is a function of shear stress, which is calculated as a product of shear rate and material viscosity. Breaking up of an agglomerate will occur only when the shear stress exceeds the strength of the particle.

Commonly used mixing procedures in the industries are dump mixing, in which all ingredients are added to the mixer at once; upside down mixing, in which solid additives are added first, followed by the polymer; seeding mixing, in which a small amount of previously well-mixed batch is added to the new batch; and sandwich mixing, in which part of the major component is added first, then minor components, followed by the rest of the major component. As a rule of thumb, hard additives are added as early as possible in the mixing cycle. Dilutes, on the other hand, are added as late in the cycle as possible.

A series of different tests can be performed with the microprocessor-controlled torque rheometer in conjunction with different sensor systems. This section discussed various tests with MIMs.

III. RELATIONSHIP BETWEEN TORQUE AND SHEAR STRESS/RPM AND SHEAR RATE

Over a limited shear rate range, the flow behavior can be governed by the power law equation:

$$\tau = m\dot{\gamma}^n \tag{1}$$

The temperature dependence of consistency is given by:

$$m = m_0 \exp(\Delta E/RT) \tag{2}$$

where exp( ) refers to "e" raised to the quantity in brackets. Due to the complex geometry of the mixer rotors, and because rotors turn at different speeds, the local wall shear rate varies throughout the chamber;[3] however, we can write:

$$\bar{\gamma} = C_1 N \tag{3}$$

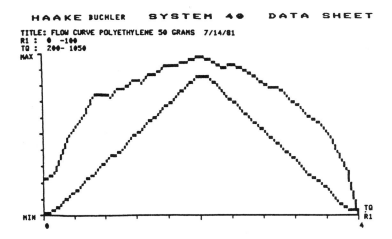

FIGURE 3. Data generated by using speed programming to test the shear sensitivity of the polymer.

$$\bar{\tau} = C_2 \hat{T} \tag{4}$$

where $\hat{T}$ = torque, N = rpm, C_1, C_2 = constants, and $\bar{\gamma}$ and $\bar{\tau}$ are the mean shear rate and mean shear stress averaged over the surface of the chamber. From the above relationship, the following expression between torque and RPM can be obtained:

$$\hat{T} = k \exp(\Delta E/RT) N^n \tag{5}$$

Calculation of the actual activation energy and the power law index are possible.

A. SHEAR SENSITIVITY

Shear sensitivity of the test material can be obtained from data generated as illustrated in Figures 3 and 4. In order to obtain the data, the power law constitutive equation is adopted.

Since torque is proportional to shear stress and RPM is proportional to shear rate, as discussed earlier, the following relationship between torque and RPM may be obtained:

$$\log(\hat{T}) = \log m' + n \log N \tag{6}$$

Therefore,

$$n = \frac{d \ln \hat{T}}{d \ln N} \tag{7}$$

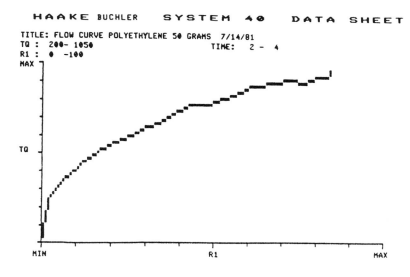

FIGURE 4. Plot of torque vs. RPM.

Figure 3 illustrates the study of polyethylene using speed programming: (0 to 80 RPM) ramp time with 2 min up and down. The data obtained are presented as a plot of torque, RPM vs. time curve. Figure 4 shows a flow curve of the polyethylene, re-plotted (torque vs. RPM) from the data shown in Figure 3. This test is used to obtain the flow behavior of the material as a function of speed.

Generally, the viscosity of a polymer decreases as temperature increases, and vice versa. Properties of the material also change depending on temperature. Knowing the temperature dependence of viscosity is important for the engineers processing compounds. Large variations of viscosity for a certain range of temperatures means that the material is thermally unstable (i.e., requires large activation energy). This kind of material must be processed with accurate temperature control of the processing equipment.

Figure 5 shows the torque vs. temperature curve obtained from the microprocessor-controlled torque rheometer. In this experiment a study of the temperature dependency of viscosity-related torque on PVC was made. These data can be applied to the Arrhenius equation

$$\ln(\eta) = \ln(A') + \frac{\Delta E}{R}\left(\frac{1}{T}\right) \tag{8}$$

Because viscosity is proportional to the ratio of torque and RPM, the relationship

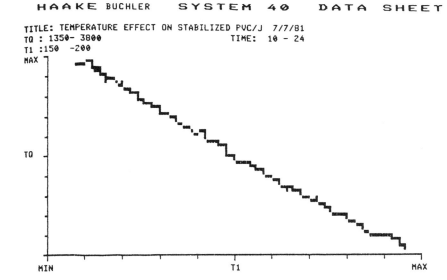

FIGURE 5.   Torque vs. temperature curve generated by torque rheometer and MIM.

$$\ln\!\left(\frac{\hat{T}}{N}\right) \,=\, \ln A' \,+\, \frac{\Delta E}{R}\left(\frac{1}{T}\right) \tag{9}$$

may be obtained.

It is often difficult to explain flow behaviors of materials obtained from the torque rheometer due to melt temperature fluctuations induced by frictional heat. Therefore, it is necessary to compensate for the temperature effect of the material. Temperature-compensated torque can be obtained from an integrated form of Arrhenius equation[4]:

$$\ln\!\left(\frac{\hat{T}}{\hat{T}'}\right) \,=\, \frac{\Delta E}{4{,}576}\left(\frac{1}{T} - \frac{1}{T'}\right) \tag{10}$$

where $\hat{T}$ = measured torque at temperature T (degrees Kelvin) and $\hat{T}'$ = calculated torque at reference temperature T'(degrees Kelvin).

Induction time is defined as the time at which a testing material begins to oxidize. Induction time of a material can be obtained from torque rheometer data. It can be seen as decreasing or increasing torque on the torque curves, depending on structural breakdown or network formation due to oxidation at constant temperature, respectively.

These data can be obtained using the temperature-compensated torque described above. Figure 6 shows data obtained from the torque rheometer utilizing temperature-compensated torque software.

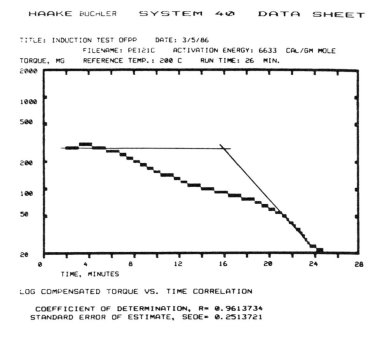

TITLE: INDUCTION TEST OFPP   DATE: 3/5/86
          FILENAME: PE121C   ACTIVATION ENERGY: 6633   CAL/GM MOLE
TORQUE, MG   REFERENCE TEMP.: 200 C   RUN TIME: 26 MIN.

LOG COMPENSATED TORQUE VS. TIME CORRELATION

COEFFICIENT OF DETERMINATION, R= 0.9613734
STANDARD ERROR OF ESTIMATE, SEOE= 0.2513721

FIGURE 6.   Typical data showing effect of temperature on degradation of polymer.

Induction time of the polymer at a temperature can be obtained by extrapolating the stable torque and sharp decreasing torque regions as shown in Figure 6. The data also can be obtained using torque curves directly obtained from the rheometer.

The unit work is defined as the work energy required to process the unit volume or unit mass of material. This can be calculated from the totalized torque, which is obtained from the rheometer directly. Totalized torque is defined as the energy required to process a certain material for a certain period of time at given conditions, and simply as the area under the torque curve. The totalized torque can be converted to work energy:

$$W_u = \frac{W_t}{V_b}$$

$$= \frac{2\pi N \int_{t_1}^{t_2} \hat{T}(t)dt}{V_b} = \frac{61.588 \times N \times TTQ}{V_b} \quad (11)$$

where a unit of unit work = $[J/cm^3]$, W_t = total work energy, $\hat{T}(t)$ = torque, N = RPM, V_b = charged sample volume, and TTQ = totalized torque.

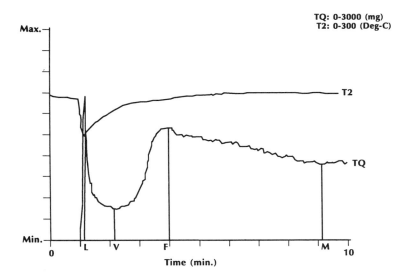

FIGURE 7. Experiment illustrating the fusion characteristics of PVC. Fusion time is defined from the loading peak (point L) to the fusion peak (point F).

B. FUSION CHARACTERISTICS

Lubricants play an important role in the processing and in the properties of the final product. Lubricants also affect the fusion of the polymer materials; i.e., internal lubricants reduce melt viscosity, while external lubricants reduce friction between the melt and the hot metal parts of the processing equipment and prevent sticking, controlling the fusion of the resin[5].

Figure 7 illustrates the fusion characteristics of PVC. The level of external lubricant used in the formulation affects fusion time between point L and point F on the curve[6]. The higher the level of external lubricant in the formulation, the longer the fusion time.

If an unnecessarily high level of external lubricant is used in the formulation, it will take a longer period of time to melt the material processing, which results in the reduction of production, an increase in energy consumption, and poor products. Meanwhile, if too low a level of external lubricant is used, the material will melt too early in the processing equipment, which may result in degradation in the final product. Therefore, selecting the optimum amount of external lubricant is a necessity for the improvement of processing and for good quality of products.

C. SELECTION OF BLOWING AGENT FOR FOAM PRODUCTS

Foam products can be produced by adding chemical blowing agents to the formulation. The blowing agents are inorganic or organic materials and decompose under heat to yield gaseous products.

P1: 0-4000 Gas Flow (SCCM)
T2: 0-300 Melt Temp (°C)

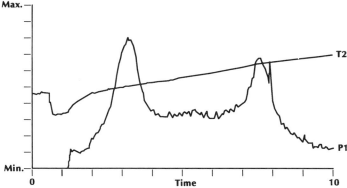

FIGURE 8. Experiment in which activation of blowing agent used in an EPDM formulation was studied as a function of temperature.

The selection of a blowing agent is very important in foam products. If the proper blowing agent is not chosen, it may react with other ingredients used in the formulation, which results in retarding the blowing agent from being activated.

The level of blowing agent used in the formulation is also important. If the level of the agent used is too high, too many unnecessary cells form in the foam product and it will collapse easily. If the level is too low, the product will not have enough cells, and so it will decrease the cushionability of the product.

Figures 8 and 9 illustrate evolution of gas from the activation of blowing agents used in rubber. Figure 8 gives information on the temperatures at which the two blowing agents begin to activate. Figure 9 provides an idea of the torque levels at which they activate.

One can study the suitability of the blowing agent chosen by the processor for formulation and determine if it reacts to any other ingredients in the formulation processing.

D. FORMULATION

Most polymers are used with additives such as fillers, lubricants, flame retardants, stabilizers, color pigments, plasticizers, impact modifiers, etc., to improve processibility, physical and mechanical properties, uniformity, flexibility, and to reduce production cost.

E. ADDITIVE INCORPORATION AND COMPOUNDING

All of the additives used in a formulation must be incorporated in the major component, and the components should be in a stable molecular

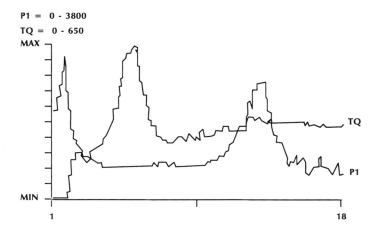

FIGURE 9.   Torque level when blowing agents in an EPDM compound are activated.

arrangement. Figure 10 is a test result for incorporation of minor components to the major component, as well as a homogeneous compound after the additives are incorporated. The test was performed with synthetic butyl rubber (SBR) rubber and preblended additives based on ASTM D 3185 (American Society for Testing and Materials, Philadelphia).

Ethylene propylene diene monomer (EPDM) rubber was loaded into the mixer and mixed for 30 s. Preblended additives (carbon black, zinc oxide, sulfur, and stearic acid) were added at 30 s. Torque values dropped sharply immediately and increased as the additives were incorporated; when the ingredients were fully incorporated, it generated the second peak, and torque values stabilized when the material was homogeneously compounded. The second peak is called the "incorporation peak".

If hard fillers are added to the polymer, torque increases sharply and generates the second peak. This can be seen when carbon black is incorporated.

The interval from the time the minor components are added to the major component to the point of the "incorporation peak" is called "incorporation time".

Cross-linkable thermoplastics, rubber, and thermoset materials change their physical properties, usually from liquid to solid (cure) via chemical reaction, heat, or catalyst during processing.

It is important for process engineers to know how long it takes to cure, how high a torque is generated, and how long it will be in stable flow. This information would help the engineers predict how the material should be processed to provide the best processibility and properties. Figure 11 illustrates the test results of the cure characteristics of rubber. Once the rubber is introduced to the heated mixer, it generates a sharp increase in torque called the

TQ 0 - 5000 [mg]

T2 0 - 300 [deg - C]

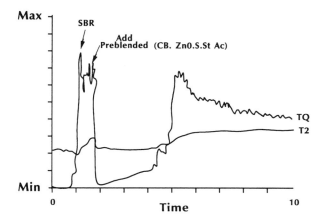

FIGURE 10. Experiment illustrating the incorporation of ingredients in a rubber compound.

T Q: 0 - 2000 [mg]

T 2: 0 - 300 [deg - c]

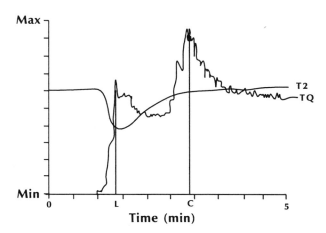

FIGURE 11. Experiment shows scorch and cure characteristics of a rubber.

''loading peak''. After it is loaded, the material begins to soften, which results in decreasing torque. It reaches minimum torque (maximum flow) and starts to scorch prior to the curing reaction. When it cures, the torque value increases and torque decreases after the curing reaction ends. The cycle time is considered to be the time from loading to cure peak.

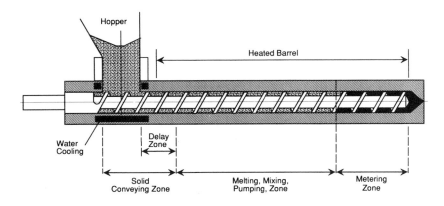

FIGURE 12.   Schematic of single screw extruder.

IV. SINGLE SCREW EXTRUSION

Solids conveying, melting, mixing, and pumping are the major functions of polymer processing extruders. The single screw extruder is the machine most widely used to perform these functions.

The plasticating extruder has three distinct regions: solids-conveying zone, transition (melting) zone, and pumping zone (shown in Figure 12). It is fed by polymer in the particulate solids form. The solids (usually in pellet or powder form) in the hopper flow by gravity into the screw channel where they are conveyed through the solids-conveying section. They are compressed by a drag-induced mechanism, then melted by a drag-induced melt removal mechanism in the transition section. In other words, melting is accomplished by heat transfer from the heated barrel surface and by mechanical shear heating.

Mixing can be carried out either in a solid or in a molten state, and is achieved through the application of shear to the material. Pumping forces the molten polymer through a die to shape the commercial product or for further processing.

Solids conveying is one of the basic functions in the screw extruder. The polymer particles in the solids-conveying zone exert an increasing force on each other as the material moves forward, and voids between the pellets are gradually reduced. As the particles move toward the transition section, they are packed closely together to reach a void-free state, and form a solid bed which slides along the helical channels.

The solids-conveying mechanism is based on the internal resistance of a solid body sliding over another (friction) generated between the plug and barrel surfaces and the screw. This type of flow is known as a drag-induced plug flow. The frictional force between the barrel surface and the solid plug is the driving force for the movement of the plug; the forces between the screw and the plug retard the motion of the plug in the forward direction.

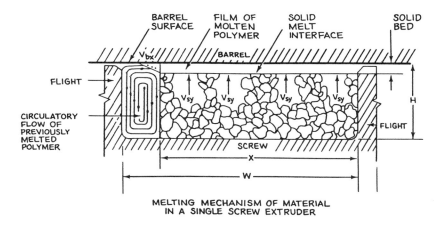

FIGURE 13. Melting mechanism of polymers in a single screw extruder.

Figure 13 illustrates the melting mechanism of material in the extruder. Melting occurs due to mechanical and thermal energy transformed into heat. The plug formed in the solids-conveying zone generates friction in contact with the heated barrel surface, and in contact with the screw. Both of these frictional processes result in frictional heat generation which raises the material temperature; this in turn exceeds the melting temperature or softening point of the polymer, and will convert the frictional drag into a viscous drag mechanism. A melt film created between the hot metal and solid bed.

As the plug moves forward, the melt portion increases and forms a melt pool, which becomes larger and larger. The conveying mechanism in this zone is one of viscous drag at the barrel surface, determined by the shear stresses in the melt film and frictional (retarding) drag on the rest of the screw and the flights[2].

Solid and molten material coexist in the melting zone of the extruder. Figure 14 illustrates the melting procedure in the single screw extruders. In this figure, the sequence of events is

1. Solids conveying section (delay zone)
2. Beginning of the transition (formation of melt film)
3. Formation of melt pool
4. Melting continues and the width of the solid bed decreases, as the channel depth continues to decrease as it progresses down the transition.
5. The solid bed break-up
6. The plastic continues down the metering section to the discharge.

An ideally designed screw gives zero solid bed profile at the end of the screw.

Melt conveying occurs in two distinctive regions. One region is downstream of the melting zone (after the completion of melting) and the other is in the melt pool, which is an extension of the solid bed profile.

FIGURE 14.   Melting progression in a single screw extruder.

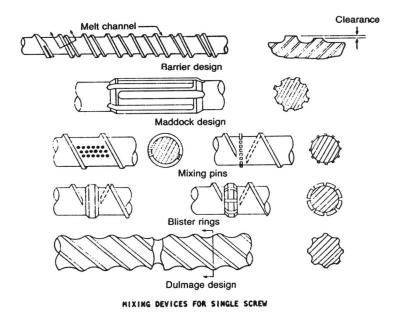

FIGURE 15.   Different mixing configurations used in single screw extrusion studies.

In the metering section or at the end of the screw, the polymer transforms totally into a melt. At this point, the solid bed profile, which is the ratio of the length of the solid bed to the screw lead length, must reach zero so that only melted polymer comes out of the extruder and die. In order to create a homogenous melt, mixing screws, barrier screws, and sometimes two stage screws are often used.

A standard single screw extruder can exhibit good mixing properties within limits. Mixing is achieved through the application of shear to the material, and constant orientation of the flow pattern. The screw can be modified to improve mixing by adding a mixing section following the metering section. The mixing section is specially designed to break up flow patterns.[10] Figure 15 shows some of the mixing devices used frequently in screw design. Mixing sections have no pumping action, and the metering section behind it pumps the melt through this section. Mixing always results in heating the melt by mechanical work, so many polymer melts would degrade. Therefore, the amount of mixing often must be limited.

It is important to select proper materials for the construction of a screw. Machine ability, material strength, abrasion resistance, and corrosion resistance are major factors in the selection of materials. Screws made of stainless steel 4140 with chrome (or nickel) plating and that are flame hardened are widely used. Screws made of duranickel or Hastelloy C are used for corrosion resistance. X-alloy barrels are considered for abrasion resistance, and Hastelloy barrels are used for corrosion resistance.

The solids-conveying section determines the effectiveness of solids feeding or conveying. The depth of the feed section should be determined based on the bulk density of the material and the strength of the remaining screw diameter to withstand the maximum torque of the drive. It is desirable to take the maximum flight depth possible while maintaining the mechanical strength of the screw, especially for low bulk density material. A flight land of one tenth of the screw diameter is normally used. If the tip of the flight is too narrow, the flight may crack during use due to weakness caused by erosional force.

The length of any zone (solids conveying, melting, melt conveying) depends on the operating conditions, the screw geometry, and the physical properties of the polymer being processed. Longer feed sections are usually desirable for a material that has a higher melting point. Long feed sections are favorable for crystalline materials because the solid bed is rigid, and compressibility is low. Shorter feed sections can be utilized for amorphous polymers, because these materials do not have a heat of fusion and have higher compressibility.

Additional flights (only in the feed section) or grooving the barrel in the feed section would be helpful for efficient solids conveying. Melting takes place and high pressures can be developed in this section. The compression ratio (ratio of feed depth to metering depth) is a commonly used term for describing the transition of the screw from the feed section to the metering section. The higher rate of compression in the transition section can increase the rate of melting due to higher pressure. The length of the transition section is affected by the length of the feed section. If the screw has a long feed section, there will be more molten polymer in the feed section; therefore, the solid polymer will be at a higher temperature by the time it reaches the transition section. Consequently, the polymer is more easily deformed, and a shorter transition section, or greater compression ratio, can be used.

In the metering section, melting continues to take place and pumping also occurs. It is ideal that the solid bed profile (amount of solids present in the screw channel at the end of this section) reaches zero. Homogenous melt can be obtained in this way. The more solids that exist near the end of the metering section, the less homogenous the extrudate. The depth of the metering section is determined by the melt viscosity of the resin. The higher the viscosity, the

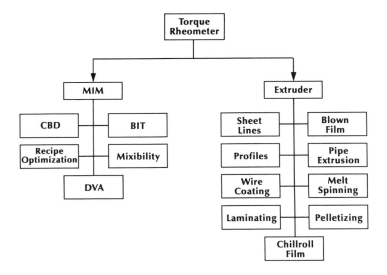

FIGURE 16. Simulations of various extrusion and miniaturized internal mixing (MIM) processes.

greater the flight depth should be in order to reduce viscous shear heating. When a deep flight depth is used, a long metering section should be employed to ensure complete melting of the material.

It is difficult to balance the first stage output with the second stage output. If the first stage delivers more than the second stage pumps through the die, the result is vent flow. If the second stage tends to pump more than the first stage delivers, the result is surging of output and pressure. This can be adjusted by controlling the feed or by valving the output. In the venting section, maintaining zero pressure drop is the most important factor. A flight depth similar to the feed depth, or approximately 15% greater than the first metering section, is usually adequate. A multiple flighted configuration can be used for the vent section to increase the amount of surface exposure and the efficiency of devolatilization. The vent section should have at least three flights, or more if possible.

A. APPLICATIONS OF SINGLE SCREW EXTRUDERS

Production simulations, rheological property measurements, and processing characteristics are discussed in this section. The applications mentioned above that use a single screw extruder are widely acknowledged in polymer industries and research organizations because the production line can be closely simulated in the laboratories.

Simulation of the extrusion process in the laboratory is one of the most important applications of the torque rheometer in conjunction with single screw extruders. Figure 16 illustrates simulations of widely used extrusion processes in the industries.

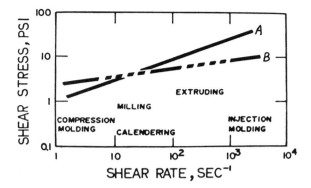

FIGURE 17. Approximate shear rate ranges for different processing equipment.

It is very important for a process engineer to know the rheological properties of a material, since the properties dominate the flow of the material in extrusion processes and also dominate physical and mechanical properties of the extrudates. Therefore, it is also important to measure the properties utilizing a similar miniaturized extruder in the laboratory, so that a process engineer knows the flow properties in the system by simulating the production line. This will enable engineers to see lot-to-lot variances of materials for better quality control of the resins.

It is desirable to know the flow properties of a material to be processed in the range of shear rates of equipment to be used. Figure 17 illustrates approximate shear rate ranges processed in different processing equipment, as well as that rheological properties measured at a lower shear rate ranges cannot be applicable to high shear rate ranges, and vice versa, because the viscosity of material A is lower than that of material B at low shear rate ranges (1 to 10/s), but it is the opposite at the high shear rate ranges (>100/s).

In order to obtain viscosity data, a slit die with three pressure transducers along the die can be used. Because the slope of pressure drop vs. length of die (L) curve in the fully developed region is used to calculate shear stress, entrance effect correction is not necessary in slit rheometry.

The viscosity of materials can be calculated as follows:

Apparent shear rate (Γ_w):

$$\Gamma_w = \frac{6Q}{wh^2} \tag{12}$$

True shear rate ($\dot{\gamma}_w^*$) with Rabinowitch corrections is:

$$\dot{\gamma}^* = \frac{\Gamma_w}{3}\left(2 + \frac{d \ln \Gamma_w}{d \ln \tau_w^*}\right) \tag{13}$$

True shear stress ($\tau_w{}^*$) is:

$$\tau_w{}^* = \frac{h}{2L} (\Delta P) \tag{14}$$

where h = opening of slot (cm), w = width of slot (cm), L = length of slit die (cm), Q = volume flow rate (cm³/s), ΔP = pressure drop (dyne/cm²).

Even though they are the same polymer materials, often they show lot-to-lot variances in processing and in product quality. The reasons the variance occurs are difficult to detect in the production line because they are dealt with en masse.

The processing characteristics can be easily tested in the laboratory by simulating the production line and analyzing the data microscopically. A miniaturized single screw extruder in conjunction with a torque rheometer can characterize the processing by measuring total shear energy (mechanical energy), specific energy, residence time, shear rate, and specific output.

B. TOTAL SHEAR ENERGY (TE)

Total shear energy (TE) is the energy introduced to the polymer material by the motor drive during processing. This is calculated from the measured torque multiplied by screw speed (N).

$$TE = Torque \times N \times 9.807 \times 10^{-3} \tag{15}$$

These data provide information about total mechanical energy required to process the material.

C. SPECIFIC ENERGY (SE)

Specific energy (SE) is defined as the energy required to process a unit mass of material. This is calculated from TE divided by the total mass flow rate.

$$SE = \frac{0.0167 \, TE}{\dot{m}} \tag{16}$$

where $\dot{m}$ = total mass flow rate. These data provide information about viscous dissipation heat built-up in the system based on screw speeds.

D. RESIDENCE TIME (t′)

Residence time is the time required for a material to reside in the extruder before it comes out of the die. It can be obtained from volume of screw (V) divided by volumetric flow rate (Q), or length of screw (L) divided by angular velocity of screw (πDN):

$$t' = \frac{V}{Q} \quad \text{or} \quad t' = \frac{L}{\pi DN} \tag{17}$$

Maintaining uniform residence time is important in achieving a quality product. If it is not maintained, the material will sometimes degrade because of excess exposure to heat.

E. SHEAR RATE ($\dot{\gamma}$)

Shear rate in the metering section of the screw can be obtained as follows:

$$\dot{\gamma} = \frac{\pi DN}{h} \tag{18}$$

where D = diameter of screw, N = RPM, and h = gap between screw and barrel.

F. SPECIFIC OUTPUT

Specific output (SO) is defined as mass flow rate per unit RPM of screw and is calculated from total mass flow rate ($\dot{m}$) divided by RPM (N) used.

$$SO = \frac{\dot{m}}{N} \tag{19}$$

This information characterizes the uniformity of solids-conveying, melting, and pumping mechanisms of the screw being used.

The following data have been obtained using fiber-graded polyproplene from different lots. Sample A is standard control material, and B and C are from different batches. The test was performed using an instrumented single screw extruder (L/D = 25, D = 3/4 in.), with a two-stage screw and capillary die (L/D = 40), in conjunction with a microprocessor-controlled torque rheometer. The temperature profile on the extruder barrel was 220°C at zone 1, 200°C at zone 2, 190°C at zone 3, and 190°C at the die. The melt temperature was measured with an exposed thermocouple, and entrance pressure was directly measured at the entrance to the capillary.

This test provides a microsopic view of the processing characteristics of the polymer, showing differences that would normally not be seen individually in the overall macroscopic view seen in the production line. Figure 18 shows normalized output (grams of material extruded per screw revolution) vs. RPM. Normalized output tends to decrease as screw speed increases, even though overall output increases as screw speed increases because of melt limitations of the material in the system. Figure 19 illustrates specific energy (energy needed to process 1 g of material) vs. RPM. The specific energy increases steadily as screw speed increases because the overall viscosity of the material in the system increases due to decreasing residence time as screw speed increases.

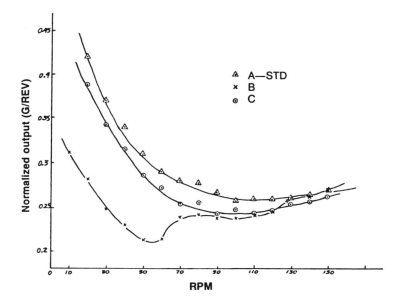

FIGURE 18.   Normalized output as a function of screw RPM.

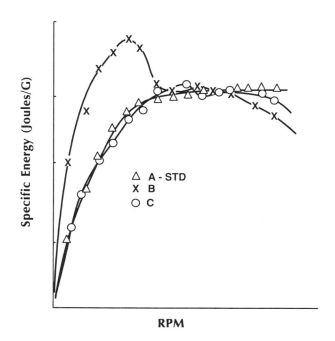

FIGURE 19.   Specific energy as a function of screw RPM.

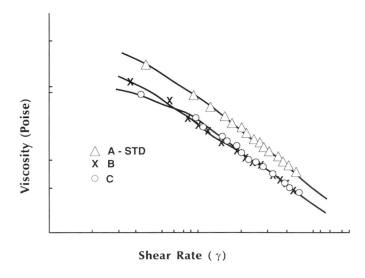

FIGURE 20.   Plot of viscosity vs. shear rate.

Figure 20 shows a plot of viscosity vs. shear rate. The viscosities of samples B and C show much similarity, but the specific output and energy still show large differences. The test results lead to the conclusion that the same material from different lots show different processing characteristics.

V. TWIN SCREW EXTRUSION

Although a single screw plasticating extruder is satisfactory for melting and extrusion, its mixing capabilities are limited. Tougher, specialized thermoplastic materials developed for new applications must be processed at conditions requiring very accurate temperature control, uniform residence time distribution, better dispersion of fillers and additives, positive venting, and surge-free extrusion of homogeneous melt for high-quality products. Twin-screw extruders are widely used for such applications, and they overcome limitations of the single-screw extruder and the intensive batch mixer. The twin-screw unit comes in a variety of configurations; they can be non-, partially, or fully intermeshing; they can be conical or cylindrical. This section discusses the basic principles of solids-conveying, melting, mixing, degassing, energy, and residence time of the counterrotating, partially intermeshing, conical twin-screw extruder. Figure 21 shows the flow exchange in partially intermeshed twin screws and in fully intermeshed twin screws.

Material feeding is important in extrusion. Starved feeding is a common practice in order to avoid drive and thrust-bearing overload, but forced feeding has applications as well. Two types of feeding devices, i.e., the force feeding (crammer feeder) and the screw-metering feeder, are commonly used to regulate and control the feed rate.

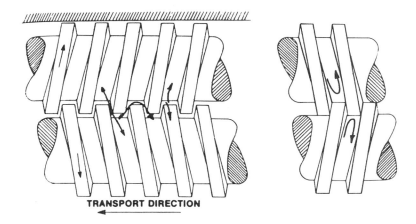

FIGURE 21. Flow interchange in a partially intermeshed twin screw (left) and in a fully intermeshed twin screw (right).

When low bulk density material (granulate or power) with low bulk weight fillers such as fumed silica or wood dust are processed, the air in the feed is forced out in the intake zone, and can impede the smooth intake of the inflowing material. Since the throughput rate of the extruder is limited by the volumetric-conveying rate, the extruder will tend to surge. Therefore, pre-compressing the feed material is required to increase the output rate, and it can be done by means of force-feeding devices at the throat.

High bulk density material with high bulk density filler, such as high-temperature engineering plastics with carbon blacks, on the other hand, cannot be force fed due to rheological and physical properties of the material. In this case, starve feeding, utilizing the gravimetric or volumetric metering feeder to control the feed rate and to avoid thrust-bearing overloading, is a common practice. The desired output rate can be obtained by maintaining optimal processing conditions and by controlling the feed rate.

Figure 22 shows the cross-section of the feed zone in a counterrotating-screw system. In order to feed as much material as possible, utilizing the largest conveying volume, the screws turn outward on top and inward at the bottom. To obtain more material conveying, a different screw geometry just underneath the hopper often uses a greater pitch and multiple thread starts. The positive displacement of such a filling zone is greater than that of the rest of the extruder. Polymer enters the screw in the intake zone, and is compressed and conveyed toward the downstream into the melting zone, where transition from solids to melt occurs.

Meanwhile, at the edge of the solids bed, a melt pool is created that draws in solid particles and softens them, and the plasticated mass is taken in by a calender gap. Inflowing unmelted polymer goes into the melt stream by kneading action in the zone in which the melting begins to accelerate the

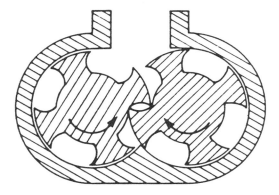

FIGURE 22.   Cross-section of the feed zone in a counterrotating screw system.

melting process. Melt can be conveyed while the trapped air can leak back. The melt will eventually reach a point at which pressure buildup commences as a result of the flow of melt back through the various leakage gaps.

In a single screw extruder, the solids bed steadily diminishes along the melting zone. In a twin screw extruder, however, a complete melting sequence can be found in each chamber where the solids bed and melt interface move from the diverging side to the converging side of the screws while the chambers move through the extruder. Janssen[11] has extensively studied the melting mechanism in the twin screws.

Figure 23 shows unwound chambers to illustrate the melting process. The melting mechanism at low backpressure is also different from that at high backpressure. At low backpressure, the melting process takes place at the barrel wall so that heat transfer plays an important role. With high backpressure, however, the fully filled length of the extruder reaches back to the melting zone, and considerable leakage flows take place. In this case, melting starts at the solids/melt interface. Therefore, headpressure affects the melting inside the system.

It is well known that compounds with two or more components should be mixed well to provide optimum properties in the final product. There are two distinct types of mixing processes: the spreading of particles over positions in space (distributive mixing), and the sharing or spreading of the available energy of a system between the particles themselves (dispersive mixing).

Distributive mixing is used for any operation employed to increase the randomness of the spatial distribution of particles without reducing their sizes.[1] This mixing depends on the orientation of flow and the total strain, which is the product of shear rate and residence time. Therefore, the more random the arrangement of flow pattern, the higher the shear rate; the longer the residence time, the better the mixing.

Dispersive mixing is similar to that of a simple mixing process, except that the nature and magnitude of forces required to rupture the particles to an

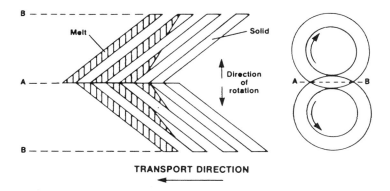

TRANSPORT DIRECTION

FIGURE 23. Unwound chamber illustrating the melting process.

ultimate size must be considered.[1] An important application of such intensive mixing is the incorporation of pigments, fillers, and other minor components into the matrix polymer. This mixing is a function of shear stress, which is the product of shear rate and material viscosity. Breaking up of an agglomerate will occur only when shear stress exceeds the strength of the particle.

Figure 24 illustrates the various leakage flows studied in depth by Janssen[11] and others. The magnitude of these leakage flows is dependent on the level of pressure differences between the chambers, but because there is communication between chambers all the way to the die, the leakage flows become a function of the die pressure.

The leakage flows between chambers, although reducing the output rate, can bring about much intensive mixing, which is of great importance in the twin screw extruder. Idealized theoretical output (Q_{th}) from a twin screw extruder can be calculated as:

$$Q_{th} = 2mNV \tag{20}$$

where m = number of thread starts per screw, N = screw speed, and V = chamber volume. In practice, the output rate (Q_{pr}) is less than theoretical throughput because of the presence of leakage flows within the extruder.[11]

$$Q_{pr} = dQ_{th} \tag{21}$$

where the empirical constant d is between 1 and 0.

Such interaction between the chambers generates high shear at points at which the screws intermesh, since the material encounters highest shear rates in this region of the machine. Figure 25 shows the different flow patterns of co-rotating and counterrotating screws. When the screws are rotating in opposite directions, material is forced and milled in the calendar gap and is subjected to extremely high shear. These high shear rates contribute to dis-

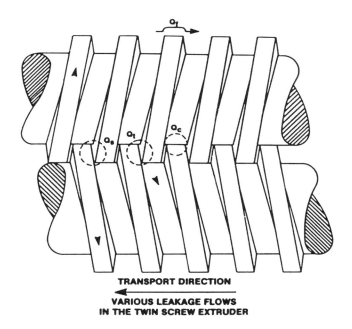

FIGURE 24. Various leakage flows in partially intermeshed twin screws.

persive mixing. Figure 26 shows pressure buildup in co- and counterrotating screws. Pressure buildup allows this material to compress and expand so that vigorous mixing can be achieved. Meanwhile, at the intermeshing point, material is also being exchanged from one screw to the other. In this way, distributive mixing is also being affected because of the rearrangement of flow patterns.

When the screws rotate in the same direction, they are essential in preventing the material from entering the gap between the screws, because as one screw tries to pull the material into the gap, the other screw pushes it out. Therefore, the material is not subject to high shear, and dispersive mixing will not be as thorough as in the counterrotating screws. A better distributive mixing, however, can be accomplished with corotating screws because of the frequent reorientation of the material as it passes from one screw to the other.

Adjustment of pressure buildup in the system with the conical twin screws is also versatile to a certain degree, simply by changing the converging section in the barrel, depending on the applications.

Polymer processing includes the introduction and some removal of mechanical and thermal energies. The temperature change of a fluid element in a flowing fluid system is determined by the sum of heat gain or loss by conduction to the element and the rate of viscous dissipation within the element. In a thermodynamic sense, the fluid element is a closed system, and the energy balance requires that its rate of increase of internal energy be equal to the rate at which heat is transferred to it and work is done on it.

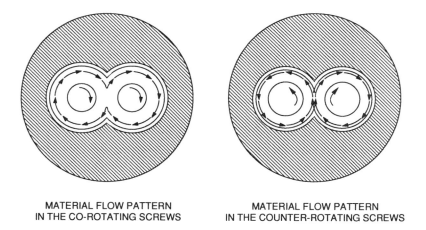

MATERIAL FLOW PATTERN
IN THE CO-ROTATING SCREWS

MATERIAL FLOW PATTERN
IN THE COUNTER-ROTATING SCREWS

FIGURE 25. Chamber-to-chamber material flow pattern in corotating (left) and counterrotating (right) twin screws.

In the extrusion system, the mechanical heat dissipation is always positive; conduction heat may be positive or negative. The specific heat and the fusion heat should be the average value of a multicomponent system.

The counterrotating, partially intermeshing, conical twin screw extruder allows the best utilization of mechanical energy because of its mixing action, which distributes the unmelted polymer into the melt stream, and because of its kneading action, which provides homogeneous melt. Because the flow streamline in the twin screws provides a high shear input, more shearing surface is generated, high pressure is developed, which can be transmitted throughout the system, and very effective heat transport can be achieved.

Practical considerations in mixing equipment usually require maintaining a relatively low temperature. These requirements originate from the temperature-sensitive nature of polymersand the large shear stress required for dispersive mixing.

Most polymer systems require venting in order to remove volatiles. This degassing, via a vent port, takes place only when volatiles migrate from the core of the material to the free surface,[12] and could be influenced by the material temperature, the pressure, and the material thickness.

These requirements can be met by conical, counterrotating twin screws. Due to intensive kneading and mixing, mechanical energy delivered by drive raises the material temperature of the melt. Pressure influence can be adjusted by changing the converging section and by changing processing parameters in order to maintain zero pressure in the vent port area. Because the screws rotate outward on top, the material is being pulled away from the center and pushed down into the barrel along the wall; material would therefore partially fill the screws in this region.

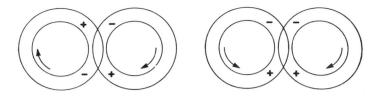

FIGURE 26. Illustrates pressure buildup in a corotating (left) and in a counterrotating (right) extruder.

Control of the residence time is of great importance for heat-sensitive polymers because prolonged exposure to high temperature may result in thermal degradation. The degree of degradation depends on the time/temperature history of the polymers; in order to contain the maximum uniform material quality, the residence time must be short and uniform. This can be achieved by constantly controlling the feed rate and by maintaining accurate processing control. Residence time distribution can be determined from the velocity profile directly, if the velocity profile is known at all locations. This mean residence time (t') equals the volume of the system (V) divided by the throughput (Q).

REFERENCES

1. **McKelvey, J.**, *Polymer Processing,* John Wiley & Sons, New York, 1962.
2. **Tadmor, Z. and C. G. Gogos,** *Principles of Polymer Processing,* John Wiley & Sons, New York, 1979.
3. **Geodrich, J. E. and Porter, R. S.,** A rheological interpretation of torque rheometer data, *Polymer Eng. Sci.,* January, 1967.
4. **Goodrich, J. E.,** Torque rheometer evaluation of poly propylene process stability, *Polymer Eng. Sci.,* 10(4), 1970.
5. **Hartitz, E. J.,** The effect of lubricants on fusion time of rigid poly (vinyl chloride), SPE 31st ANTEC, May 1973.
6. **Marx, F. M.,** A torque rheometer fusion-point test for evaluating polyvinyl chloride extrudability, *Western Electric Eng.,* 11(3), October 1968.
7. **King, L. F. and Noel, F.,** Thermal stability of poly (vinyl chloride) during processing, *SPE Vinyl Tech. Newsl.,* 6(2), December 1968.
8. **Gilfillan, E. G. and O'Leary, P. G.,** Accelerated test for processing stability of poly propylene, SPE 26th ANTEC, May 1968.
9. **Hatt, B. W.,** Formulation for Rigid PVC Using Multivariate Analysis, Elkem Chemical, Ltd., England.
10. **Cheng, C. Y.,** Extruder screw design for compounding, *Plastic Compounding,* March/April, 1981.
11. **Janssen, L.,** *Twin Screw Extrusion,* Elsevier, Amsterdam, 1978.
12. **Cheng, C. Y.,** High speed mixing and extrusion using the transfer mix, SPE 36th ANTEC, April 1976.

Chapter 7

RHEOLOGY AND INTRODUCTION TO POLYMER PROCESSING

I. THERMAL HISTORY

Temperature, pressure, time, stress, and flow geometry are the physical features that determine rheology. Even in simple flows, such as capillary extrusion, the temperature, pressure and stress history may be complex. Provided we understand what those factors are and how they influence the flow, one can often ignore the complexity by limiting analysis to a pragmatic consideration of apparent viscosity. The same simplification cannot be presumed where flows of complex geometry are considered; e.g., in a processing operation we are concerned with shaping a material and the "shape" of the flow is the most important physical factor.

As heat is supplied to a polymer the molecules vibrate more rapidly and mobility is increased. Viscosity curves for a given polymer that melt at different temperatures are approximately superposable by shift at constant stress. The dependence of viscosity on temperature is itself temperature dependent, being greater at low temperature. The elastic modulus of polymer melts is usually very much less sensitive to temperature than viscosity, as shown in Figure 1.

Because of the non-Newtonian character of polymer melts the superposition of flow curves at different temperatures by a shift at constant stress means that no corresponding superposition occurs at constant shear rate. For a shear thinning material, the dependence of viscosity on temperature appears to decrease at high shear rate, as illustrated in Figure 2. The following relationship can be deduced from experimental observation:

$$\left(\frac{\text{Viscosity} \ @T_1{}^\circ C}{\text{Viscosity} \ @T_2{}^\circ C}\right)_{\text{Rate}} = \left(\frac{\text{Viscosity} \ @T_1{}^\circ C}{\text{Viscosity} \ @T_2{}^\circ C}\right)^n_{\text{Stress}} \qquad (1)$$

where shear stress is proportional to shear rate raised to the nth power.

In rheological studies on the influence of temperature on viscosity, it is a proper precaution to check that exposure to high temperature has not produced chemical alteration of the sample. The most convenient way to do this is to measure the viscosity at a low temperature at which the sample is known to be stable, both before and after the high temperature experiment.

In polymer processing, high temperature is a concern. First, it is expensive in that energy is required to raise a material to a high temperature. Second, the time necessary to cool a material to a form-stable product is time expended. High temperatures are also to be avoided where these may

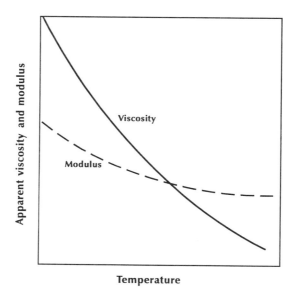

FIGURE 1. Temperature dependence of elastic modulus and viscosity at low shear.

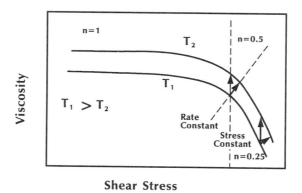

FIGURE 2. Viscosity dependence on temperatures.

lead to decomposition of the polymer. Thus, in practice, we usually seek to process at the lowest possible temperature, but strangely the lowest temperatures may often be obtained by a judicious increase in the heat by high work input. Thus, if the material can be softened by heat input early in the process, excessive and nonuniform heat generation can sometimes be avoided at a later stage.

TABLE 1
Experimental Methods to Study the Influence of Pressure on Viscosity

Technique		Advantages	Problems
Rotational Pressurized concentric cylinder	Torque detection / Shearing region / Pressure	Torque, and thus viscosity, measurement independent of pressurizing system	Limited by heat generation at high shear rates Sealing problems with low viscosities
Extrusion Pressurized capillary rheometer	P → P+ΔP	Direct descendent of capillary rheometer	Friction losses; viscosity measurement derived from small difference between two large pressures
Double-die method	P+ΔP → P	Simple modification of capillary rheometer	Pressure dependent on flow rate and die geometry
Nonlinear Bagley plot	P vs L/R	Data available from standard capillary rheometry	Very high precision of measurement required together with sophisticated interpretation

II. EFFECT OF PRESSURE

In contrast to thermal history, pressure reduces both free volume and molecular mobility, thus resulting in an increase in viscosity. Table 1 provides a summary of experimental techniques for studying the effects of pressure on viscosity.

Since the influence of pressure on viscosity is qualitatively similar but opposite in sign to that of temperature, a suitable way of representing that dependence is to describe a temperature/pressure equivalence so that for engineering purposes, pressure may be considered as a negative temperature. The application of pressure, ΔP, increases the viscosity; then, if ΔT is the temperature rise necessary to bring the melt back to its original viscosity, we may define this relationship as:

$$\left(\frac{\Delta T}{\Delta P}\right)_\eta \quad ; \quad °C/N - m^{-2} \tag{2}$$

This function has the appearance of a thermodynamic function and can be compared to the isoentropic function

$$\frac{\delta T}{\delta P}^s \quad ; \quad °C/N - m^{-2} \tag{3}$$

This is the instantaneous temperature rise resulting from the application of pressure which is a function that can be measured in a capillary rheometer.

The close correlation between these functions suggests that if no direct measure of the influence of pressure on viscosity is available, then the thermodynamic function may be used as a guide. The correlation also suggests a relation between viscosity and entropy (i.e., the degree of disorder).

One can readily recognize the dependence of viscosity on temperature during polymer melt processing, and that temperature control within 1 or 2°C is highly desirable. Most theoretical texts assume that liquids are incompressible, providing a convenient excuse for neglecting the influence of pressure. However, as 1000 atm (a common pressure in polymer processing) has as much effect on viscosity as a reduction of 50°C in temperature, this is one simplification that cannot be ignored.

In polymer processing the combination of high pressure and low temperature will tend to promote the crystallization of some materials, such that in some cases the harder fraction deforms, causing the material to flow less.

III. TIME HISTORY

In the interpretation of rheological properties using the apparent Maxwell concept measurements are usually made during steady flow. This time scale may sometimes be long relative to the time scale of real processing operations where, for example, an injection molding may be filled in <1 s or the passage through a die may be on the order of only milliseconds. In many processing operations the time scale, including melting, conveying, shaping, and solidifying, may be <100 s. Laboratory rheological experiments may take as much as 1000 s to reach equilibrium. For a complete appreciation of the rheology of a polymer melt, and particularly how that rheology interacts with features such as orientation in moldings, it is essential that the dependence of that rheology on time scales between 10^{-3} and 10^3 be understood.

Evaluation of creep experiments under constant stress before steady flow is established (refer to Figure 3) enables the interpretation of time-dependent apparent Maxwell parameters:

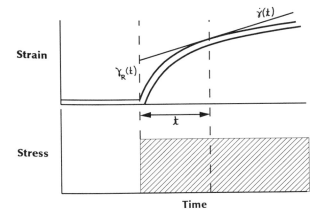

FIGURE 3. Evaluation of creep under constant stress.

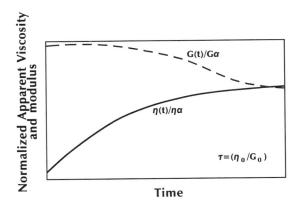

FIGURE 4. Time-dependent apparent Maxwell parameters.

$$\eta(t) \ = \ \sigma_S/\dot{\gamma}(t) \tag{4}$$

$$G(t) \ = \ \sigma_S/\gamma_R(t) \tag{5}$$

When plotted as a function of time these parameters commonly take the form shown in Figure 4.

In the region in which stress and strain are proportional, such experiments may be supplemented by an evaluation of the frequency-dependent rheological properties by assuming an inverse relationship between angular frequency and time. Within the linear region the time-dependent response is independent of stress and geometry of deformation, as shown in Figure 5.

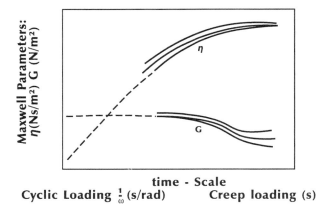

time - Scale
Cyclic Loading $\frac{1}{\omega}$ (s/rad)   Creep loading (s)

FIGURE 5.   Comparison of steady and oscillatory shear flow.

The deduction of time-dependent response in the nonlinear region is less clear. Perhaps the most instructive experiments are those in which a dynamic measurement is superimposed onto a steady flow. It is also possible to deduce the nonlinear elastico-viscous response from dynamic experiments at high strain amplitude, in which the degree of nonlinearity depends on the maximum shear stress in the material, which itself depends on the maximum shear rate (being the product of angular frequency and strain amplitude). Using a series of measurements at different strain amplitudes and taking cross-plots at constant maximum strain rate, the dynamic, or time-dependent response in the nonlinear region may be deduced. Results from both these dynamic approaches, and from creep experiments in the shear thinning region in simple shear flows, are consistent with the view that in the nonlinear region, the long time scale response is truncated.

In polymer processing a major factor reflecting the influence of the time dependency of rheology on the end-product properties is the memory that the polymer has of its processing history. The time scale for which a material has a memory is conveniently described by the natural time of the material, being the ratio between the viscosity and modulus in steady flow. By comparing the time scale of a material to the time scale of a process, the Deborah number, N_{DEB} becomes significant:

$$N_{DEB} = \frac{\text{Material time constant}}{\text{Time scale of process}} \tag{6}$$

When $N_{DEB} > 1$, the process is dominantly elastic, and when $N_{DEB} < 1$ the process is essentially viscous. In making use of this concept in process analysis it is necessary to note that because the viscosity and modulus of a material themselves depend on stress, so does the characteristic time of the material.

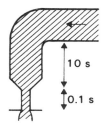

FIGURE 6. Flow around a corner.

If we consider a flow with a history of a low stress deformation around a corner followed by a high shear deformation in a die, as shown in Figure 6, one must consider two processes:

1. Flow from the bend to the die, wherein the time scale of the process may be on the order of 10 s. For low stresses the viscosity might be on the order of 10^5 Ns/m^2, with a modulus of 10^3 N/m^2. The viscosity-modulus ratio would be around 100 s, thus giving a Deborah number of about 10.
2. Flow in the die, wherein the time scale of the process is on the order of 0.1 s. At high shear stress, the viscosity might be 10^3 Ns/m^2 and modulus 10^5 N/m^2 (or viscosity/modulus = 0.01 s). Hence, $N_{DEB} = 0.1$.

In this case the remote small disturbance in the streamlines as the melt flows around the bend will be remembered at the die exit, causing the extrudate to curl, even though the melt has subsequently passed through a very large and dominantly viscous flow in the die lips. In general it is the longest time scales with which one must be concerned when considering memory effects in processing.

IV. STRESS AND DEFORMATION GEOMETRY

The most significant physical variable influencing rheology is stress. Stress may vary by four orders of magnitude during normal or conventional processing operations. Polymer melts rupture under tensile stresses on the order of 10^6 N/m^2.

A simple method of measuring rupture stress is to carry out a constant-force extension experiment. As the sample draws down, its cross-sectional area decreases and the stress increases until the sample breaks. If the breakage is superficially brittle (refer to Figure 7), the stress at rupture may be deduced as the ratio of force/area at break. Ductile, or necking failure, which is often an indication of tension-thinning behavior, must be treated with caution, and a transition to superficially brittle failure can sometimes be achieved by increasing the rate at which the experiment is conducted.

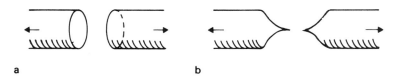

FIGURE 7.   (a) Superficially brittle failure and (b) ductile or necking failure.

Using constant-force extensions and other techniques, the rupture stress for melts can be observed to decrease for polymers of higher molecular weight and greater degrees of long chain branching. The surprising result that melts of high molecular weight are more superficially brittle than those of low molecular weight contrasts strongly with the solid state properties and possibly reflects that melt rupture is a phenomenon of stress concentration which may be expected to be more severe in highly entangled systems.

A common approach is to consider converging flow as a dominantly extensional flow. Hence, the onset of non-laminar flow (when the streamlines become discontinuous) can be used as an indication of rupture, a wider range of rupture. This viewpoint suggests that rupture phenomenon is only weakly dependent on temperature and possibly influenced by pressure.

There are a limited number of options available to increase the rupture stress of polymers. The most dramatic effects are observed in polyvinyl chloride (PVC) where the addition of acrylic processing aids greatly increases both the extensibility and the rupture stress without altering the stress/strain response of the material. The major variable open to other materials is reduction of the heterogeneity of the system by extracting volatiles and eliminating possible nuclei for failure. In branched polymers intensive shear history enhances the drawing characteristic. Also, rupture stress can be increased by a very small degree of crystallization. The presence of foreign bodies, such as dust particles, greatly increases the tendency to rupture; rupture during practical processing is nearly always associated with such contaminants.

A possible general explanation of rupture behavior is that it is associated with faults in the network on a scale of about 100 nm. Provided the tensile stress is low enough, surface tension suffices to keep such faults stable. Homogenization, improved gelation in polymers through the use of processing aids, and the application of hydrostatic pressure all help to reduce the fault size and thus to improve the rupture characteristic of melts.

Rupture is the ultimate method of failure during processing, both in thin sections such as films and fibers due to drawing stresses, and in thick sections where solidification stresses cause cavitation. However, in both extremes rupture during normal processing is most commonly an intermittent effect associated with stress concentrations. The key to minimizing or avoiding rupture is most readily achieved by avoiding notches both in the material and in the flow path. Rupture is also the critical point in the initiation of foaming.

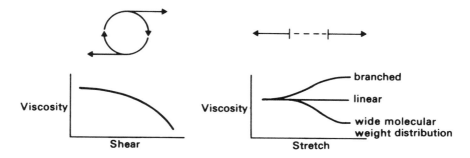

FIGURE 8. Diagram distinguishing between rotational and irrotational flows.

When an elastic band is stretched, at low stress the deformation is proportional to stress and at high stress the band stiffens rapidly toward a limiting extensibility. The same qualitative response is found in polymer melts. For polymer melts the bulk modulus is some five orders of magnitude greater than the shear modulus, so that for practical purposes, Poisson's ratio may be taken as 1/2. The bulk modulus increases with hydrostatic stress according to:

$$K_p = K_o + 9P \tag{7}$$

where K_p is the bulk modulus at pressure P above atmospheric and K_o is the bulk modulus at atmospheric pressure.

The shear and tensile module also increase with stress and at a given stress level, the tensile modulus is three times the shear modulus. For most simple polymers the melt appears to have a limiting elastic response of about six units of shear or two units of extension. The conclusion of limited elastic response allows simplification in the interpretation of very large deformations as essentially viscous flows. In practical processing, elastic deformations may play an important role in stabilizing free surface flows. The faster a deformation process is carried out the more elastic it will appear. Melt elasticity is also manifested in the orientation in products. More elastic materials will often display a greater degree of frozen-in orientation, and thus a greater anisotropy of properties and tendency to give warped moldings.

As noted earlier, the so-called spaghetti model describes the qualitative response of melts to both shear and extensional stress. If a bowlful of spaghetti is subjected to a shear (rotational) field, by digging in a fork and twisting the spaghetti around a fork, then it is possible to disentangle the elements and easily move the spaghetti from the plate to the mouth. However, if the spaghetti is subjected to an extensional (irrotational) flow by simply digging in the fork and pulling, one ends up with nothing. Thus, in shearing flows, the greater the stress, the lower the viscosity, while in extensional flows stress does not thin the material to the same extent, and indeed may even cause it to stiffen. This concept is illustrated in Figure 8.

In polymer processing the reduction of viscosity by shearing is often an essential prerequisite of attaining easy molding. The tension-stiffening response of some polymers can play an important role in stabilizing processes such as film blowing.

Most descriptions of rheological properties assume an equilibrium state as the starting point, but in practical processing situations, material response at any stage of the process may be influenced by the previous thermomechanical history of the process and even by its history in earlier compounding operations. Intense working, producing high shear, will usually lead to a reduction in viscosity and also a decrease in the elastic response. Such effects, which may at first appear to be the result of degrading the polymer, are reversed by cooking the melt.

High shear history can also significantly improve the appearance of a product, especially with respect to those properties such as gloss, the loss of which may be attributed to inhomogeneity. Studies of rheology at equilibrium reflect the equilibrium properties of the melt. These equilibrium properties may be significantly improved by mechanical working. Mechanical working may be the most effective route to making otherwise unprocessable polymers processable.

We now turn attention to deformation geometry considerations. There are three geometries of deformation: bulk, simple shear and extension. When polymers are subjected to hydrostatic tension or compression they change in volume. If the stress is a hydrostatic tension and it exceeds the rupture strength, the melt will cavitate. Bulk modulus decreases with increasing temperature and increases with hydrostatic pressure. It is independent of such factors as molecular weight and molecular weight distribution (MWD). The response to a change in pressure is very rapid. Bulk deformation and bulk stresses are significant at several different stages of processing. During melting, the specific volume may increase by up to 30% and if the material is constrained, very high stresses will be generated or the melting process may not occur at the normal temperature. During flow, small volume changes accompany pressure changes. The final stage of processing involves solidification, leading again to density changes. Because solidification inevitably occurs from the outside in, the outer shell of a molding solidifies first and the stresses produced by the solidification of the inside may lead to warping, most commonly apparent as frozen-in stress or cavitation.

An important class of processing operation involves foaming, when 50% or more expansion may typically occur because of gas evolution within the sample. This increase in volume can be used to complete the filling of a mold. The properties of such foams depend on the rate of gas evolution, the way in which the foam is nucleated, and the extensional rheology of the melt.

In a tensile, or stretching flow, material is drawn from one cross-sectional area to another. Figure 9 illustrates simple elongational flow. Such flows dominate the fiber, film, blow molding, and vacuum-forming processes and

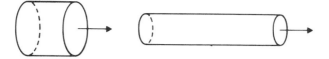

FIGURE 9. Elongational flow.

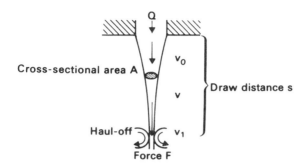

FIGURE 10. Stress variation in simple drawing of an extrudate.

those flows are basically classified as "free surface flows". The rheology in elongational flow is qualitatively different from that under simple shear, especially with respect to the dependence of viscosity on stress.

During simple drawing of an extrudate there is a very close approximation to simple extension, the principal complicating factor being that the stress varies along the line, as illustrated in Figure 10.

In addition to such obvious stretching flows, significant tensile deformations can be detected during constrained flows whenever the streamlines converge or diverge, for example, in the entry and exit regions of dies, in calenders, and in injection molding. Refer to Figure 11 for examples of tensile deformation in constrained flows.

For simple shearing flow not only does a deformation occur, but also a rotation. This rotational component leads, in polymers, to the rheological phenomena of normal stress effects and shear thinning. With practical processes we can deduce two major classes of simple shear flow: flow between relatively moving surfaces, and flow induced by a pressure gradient.

These are two important differences between such flows. First, the question of the interface: in flows between relatively moving surfaces there is the possibility of failure of adhesion between the melt and the metal surface unless there is also a sufficient hydrostatic pressure component present; in pressure-driven flows that pressure to promote adhesion is present, but if the adhesion should break down, certain classes of nonlaminar flow may be triggered. The second difference is in heat generation and exchange: in pressure-driven flows, the shear and thus heat generation takes place close to the walls where heat

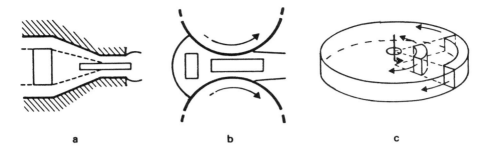

FIGURE 11. Examples of tensile deformation in constrained flows: (a) tapering section of extrusion die; (b) calendering; and (c) injection molding.

transfer away from the melt is facilitated; in flows between relatively moving surfaces the shear is homogeneous so that heat generation is uniform across the section. To be extracted, that heat must first traverse the poorly conducting melt. Flows between relatively moving surfaces dominate screw extrusion, while pressure-driven flows are most commonly found in dies and molds.

As there are important tensile contributions during practical constrained flows, superficial stretching flows may also be dominated by shear. Thus, in the squeezing of a thin film in which the application of stress appears to be extensional, the flow has a large shearing component.

V. FLOWS THROUGH COMPLEX GEOMETRY

The two types of simple shearing flow noted are pressure driven and that between relatively moving surfaces. Such flows are frequently superimposed, for example, in the barrel of a screw extruder.

To illustrate how such flows may be considered, examine a pressure-driven flow through an annulus in which the mandrel is rotated, as shown in Figure 12. Defining Q as the volume flow rate and Ω is the rate of rotation, then

$$\dot{\gamma}_{ex} = 6Q/2\pi RH^2 \tag{8}$$

$$\dot{\gamma}_{rot} = 2\pi r\Omega/H \tag{9}$$

$$\dot{\gamma}_{tot} = (\dot{\gamma}_{ex}^2 + \dot{\gamma}_{rot}^2)^{1/2} \tag{10}$$

where $\dot{\gamma}_{ex}$, $\dot{\gamma}_{rot}$, $\dot{\gamma}_{tot}$ are the extrusion, rotational, and total shear rates, respectively.

The effective viscosity in the system is defined by the total shear rate: if the material is strongly shear thinning, the addition of a rotational shear component to a pressure-driven flow may ease the flow considerably. Defining

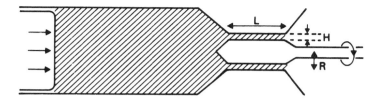

FIGURE 12. Pressure-driven flow through an annulus in which the mandrel is rotated.

ΔP as the pressure drop required to maintain extrusion, then

$$\Delta P = \frac{2L}{H} (\dot{\gamma}_{ex} \times \eta_a) \tag{11}$$

where η_a is the apparent viscosity corresponding to the total shear rate in the system.

Most capillary rheometry experiments are carried out in dies of circular cross-section, although occasionally slot dies are used. Like the simple capillary, the pressure-driven flow through an infinitely wide slot die, or annular gap, may be readily analyzed. In practice, however, many flow channels do not have such ideal cross-section.

Complex channels may be described by the following parameters: L, length of the channel land; A, cross-sectional area; 2a, length of the minor axis of the included ellipse of largest area; 2b, length of the major axis of the included ellipse of largest area; and 2c, longest chord through the axis of the die.

From these parameters the following dimensionless ratios can be defined:

$$X = \frac{a}{b} \leqslant 1 \tag{12}$$

$$Y = \frac{c}{b} \geqslant 1 \tag{13}$$

where for a capillary, $X = 1$ and $Y = 1$, and for a slot die, $X = 0$ and $Y = 1$. Using these parameters we may define a set of working relationships for pressure-driven flows which satisfy the two flows (capillary and slot). In deducing these relationships we note

	Capillary	**Slot**
Shear rate	4Q/aA	3Q/aA
Stress	$(P_L - P_O)a/2L$	$(P_L - P_O)a/L$

The working shear rate is

$$\dot{\gamma} = (3 + X)Y^{1/2} Q/aA \tag{14}$$

The working stress is

$$\sigma_S = (P_L - P_O)aY^{1/2}/L(1 + X^{1/2}) \tag{15}$$

The pressure drop through dies of zero length is

$$P_O = 4(1 + X^{1/2})^{1/2} \dot{\gamma}(\eta\lambda)^{1/2}/3(n + 1) \tag{16}$$

The swell ratio from long dies:

$$B_a = B_b = B_c = 1.0 \quad , \quad \text{for swell ratio } \gamma_R < 0.8 \tag{17}$$

and for long dies

$$B_a B_b = 1 + 0.4(1 + X/2)(\gamma_R - 0.8) \tag{18}$$

$$B_a = B_b^{(2-X)} \tag{19}$$

$$(B_c^Y) = B_b \tag{20}$$

where P_L is the pressure drop in a long die and P_O is the pressure drop in a die of zero length, B signifies the ratio of extrudate to die dimension, η is the viscosity in simple shear, λ is the viscosity in extension, n is the power law index in the equation $\sigma_S \alpha \dot{\gamma}^n$, and γ_R is the recoverable shear. The swell ratio for dies of zero length is

$$B_a B_b = \exp\{\epsilon_R^{(3+X)/4}\} \tag{21}$$

$$B_a = B_b^{(2-X)} \tag{22}$$

$$(B_c^Y) = B_b \tag{23}$$

where ϵ_R is the recoverable extension.

In converging flows (e.g., the tapered entry of a die) an extensional flow is superimposed on a shearing flows. Such flows are common in practical processing. When two shearing flows are superimposed, there is a strong interaction. A quantitative evaluation of the strain rate histories shows that at the wall of the die, where the velocity is zero, the shear strain rate has its maximum and the extensional strain rate is zero, while the maximum extensional strain rate occurs along the center line where the velocity is at its maximum and the shear strain rate is zero. In practice, it appears safe to ignore the interaction between the flows and compute the flow as determined by the addition of the shear and extensional components.

When the angle of convergence is narrow ($<10°$), shear is the dominant flow mechanism, and for the purpose of estimating the relationship between flow rate and pressure drop, the stretching flow component may be ignored. At about 45° convergence the extensional and shear components are approximately equal. If the angle is above 45° it is likely that the polymer will form its own convergence pattern by establishing recirculating regions in the corners of the die.

In designing the optimum die a balance must be struck between too much and too little streamlining. If the die has inadequate streamlining, the stretching rate may exceed that which the melt can accommodate without rupture, the streamline pattern will break down, and the extrudate will become distorted. At the other extreme, if the convergence angle is too narrow, the die becomes restricted so that the pressure requirement is unacceptably high. Thus, we require only that taper which will allow rupture to be avoided.

If the difficulties of making direct measurement of rheology in simple extension have delayed a proper appreciation of such flows, these difficulties are small in comparison to those encountered during the study of biaxial extension. Measurements made on very viscous liquids indicate that at low stress the viscosity and modulus are related to the simple elongational and shear response by the factors described in classical mechanics texts. In contrast, at high stress the response is very much less clear, although some indication of the elastic response may be deduced from studies on rubbery or crystalline solids. These suggest that the limiting elastic response in biaxial extension is the same as in simple extension.

Biaxial extensional flows play a dominant role in film blowing, blow molding, and vacuum forming. Like other stretching flows, they are rarely limited by considerations of force; indeed, melts of higher viscosity are usually preferred for such applications since, being more elastic, they produce more orientation in the final product and biaxial orientation is often a highly desirable feature. The major limitation in complex stretching flows is instability.

Chapter 8

POLYMER PROCESSING OPERATIONS

I. INTRODUCTION

This chapter provides an overview of conventional polymer processing operations. An appreciation of the equipment and dynamics of processing operations is needed in order to relate rheological properties of materials to handling and forming operations. A balance must be gained between the rheological properties and conditions under which polymers are processed and formed in order to achieve desired end-performance properties of the finished article. Often the limitations of specific process equipment or conditions over which they are operated are ignored in product design. This oversight more often than not leads to the failure of a new product in the marketplace. In other words, a polymer may meet all the performance requirements required in the final formed article and even exceed customer expectations, but if it cannot be processed in an efficient and economic fashion, not a single pound will be sold. By gaining an appreciation of large-scale processing equipment, an appreciation for the apparent processing rheology can be achieved. This knowledge helps to establish both theoretical and analytical approaches to relating performance to controlled or laboratory scale rheological characterization studies, which ultimately guide the product design specialist in the molecular design of the product. Covered in this chapter are the operations of mixing, extrusion, calendering, and molding operations. Most of these discussions will center around the author's experiences with elastomers and rubbers.

II. MIXING OPERATIONS

In rubber mixing, the objective is to produce a compound with its ingredients thoroughly incorporated and dispersed so that it will process easily in the subsequent forming operations, cure efficiently, and develop the necessary properties for end-use, all with the minimum expenditure of machine time and energy. The properties pertinent to subsequent operations are viscosity, dispersion, scorch stability, and cure rate. In order to mix efficiently, one must direct attention to raw materials, mixing procedures, mixing equipment, and quality control principles.

Among the most common apparatus used in this operation is the internal batch mixer. This machine has the ability to exert a high localized shear stress to the material being mixed (a nip-action) and a lower shear rate stirring (a homogenizing action). The effectiveness of dispersive mixing results from the combination of high shear stress and large shear deformation.

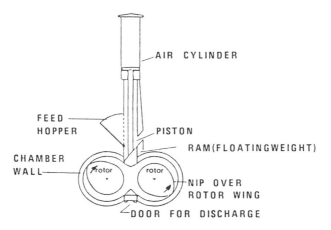

FIGURE 1.   Basic features of an internal rubber mixer. Arrows indicate pumping action.

There are two basic designs of rotor in internal mixers: nonintermeshing (e.g., Banbury, Bolling, and Werner-Pfleiderer types) and intermeshing (e.g., Inter-mix and Werner-Pfleiderer types). Typical rotor designs are illustrated in Figures 1 through 3. Intermeshing rotors provide superior heat transfer and are therefore better for heat-sensitive compounds with long mixing cycles. However, in general, rotor design has little effect on internal mixer efficiency. This is probably a result of the importance of elongational flow in the mixing process, elongational flow being the result of converging flow lines irrespective of rotor design.

Internal mixer design is a compromise, based on experience, to best accommodate the wide range of compounds typical of a manufacturing operation involving mixed product. In rubber mixing, the Banbury mixer is perhaps the most widely used. Basically this mixer consists of a completely enclosed mixing chamber in which two spiral-shaped rotors operate, a hopper at the top to receive compounding ingredients for mixing and a door at the bottom for discharging the mixed batch of compound. The rotors are driven by an electric motor while pressure is applied from the top by a plunger or ram. The two rotors subject the compound to a certain amount of shear by revolving in opposite directions and at a slightly different speed. The bulk of the shearing action, however, occurs between the rotors and the chamber wall. Water or steam is usually circulated through the hollow rotors and the chamber walls to provide cooling or heating. At the specified mixing time or temperature, the compound is discharged onto a two-roll mill where the material is sheeted off to auxiliary equipment, such as a slab cooling system.

The Bolling mixer is another design in which a ram pushes down toward the mixing chamber and the ingredients are forced between helically fluted rotors. As in the Banbury mixer, the bulk of shearing action occurs between the rotors and the chamber wall. A so-called spiral flow arrangement inside

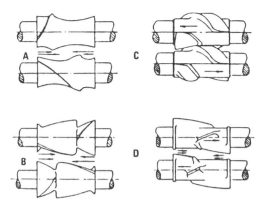

FIGURE 2. Examples of rotor designs. (A) Banbury two-wing; (B) Banbury four-wing; (C) Shaw Intermix three-wing; and (D) Werner & Pfleiderer four-wing.

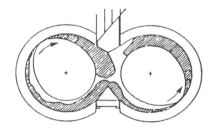

FIGURE 3. Flow lines and filling configuration of an internal rubber mixer.

the shell of the chamber wall is designed for circulation of steam to provide heat around the shell through baffles cast into the shell liner. Separate channels running through the shell liner provide water for cooling.

Mixing mills are another common apparatus used in rubber mixing operations. An open two-roll mill consists of two parallel, horizontal rolls rotating in opposite directions. The rotation of the rolls pulls the ingredients through the nip (or bite), which is the clearance between the rolls. The remaining surface of the roll is used as a means of transportation for returning the stock to the nip for further mixing. The back roll is usually rotating faster than the front roll by a ratio called the "friction ratio". Most of the work is done on the slow front roll during the incorporation of ingredients. Cold or hot water, steam, or hot oil may be circulated through the hollow rolls to control the temperature of material coming into direct contact with roll surfaces during mixing operations. The use of open two-roll mills for mixing (as distinct from finishing) is declining, at least in part because of safety and environmental regulations.

Other operating arrangements common to rubber processing are continuous internal mixers and mixing extruders. A masterbatch line for tire compounds that consists of an internal mixer that dumps the compound into the hopper of a continuous mixer or mixing extruder, can be considered continuous. The hopper holds two to three batches at a time and the output from the pelletizer or roller die is essentially continuous. The continuous mixer or mixing extruder homogenizes the product from the batch mixer and therefore allows shorter mixing cycles. In contrast, the Farrel Continuous Mixer (FCM) is a true internal mixer, with rotors and mixing action similar to the Banbury mixer. The machine does not work on the extruder principle. Raw materials are fed automatically from feed hoppers into the FCM, in which the first section of the rotor acts as a screw conveyor, transferring the ingredients to the mixing section. The action within this mixing section is similar to that within a Banbury mixer, incorporating intensive shear of material between rotor and chamber wall and a rolling action of the material itself. Interchange of material between the two bores of the mixing section is an inherent feature of the design of the rotors. The amount and quality of mixing is flexible and can be controlled by altering the speed, feed rates, and orifice opening. As the feed screw is constantly starved and the mixing action is rotary, there is little thrust or extruding action involved. Production rates and temperature are controlled by the rotor speed and the discharge orifice.

Another example of a continuous internal mixer is the Stewart-Bolling Mixtrumat apparatus. This is a high-intensity continuous device that consists of a combined twin rotor mixing station and extruder. The mixing section provides a working action of the type described earlier for internal batch mixers. The extruder section provides the coordinated means of handling the discharge from the mixing section and the conversion of it to a usable form, such as strip, pellet or, in some cases, a finished shape. Both sections of the Mixtrumat are integral, permitting direct transfer from mixer to extruder without heat loss. At the transfer section there is a reduction in pressure and a vacuum port can be installed to remove volatiles.

Powdered or particulate rubber can be processed in internal batch mixers. The two systems include the Bayer Sikoplast Screw/Hopper and the Farrel M.V.X., which are shown schematically in Figures 4 and 5. It is important to note that there are several steps in the conversion of elastomers and other compounding ingredients into rubber compounds: receipt of raw materials, testing, storage, weighing, feeding, mixing, batch-off, cooling, testing, storage, and dispatch. These stages can be divided into three groups: (1) material flow to the mixer, (2) mixing, and (3) material flow away from the mixer.

In performing compounding, mixing involves three simultaneous processes, namely, simple mixing, laminar mixing, and dispersive mixing. The relative importance of each depends on the particular compound formulation (in terms of the attraction between the particles of solid additives and the flow properties of the elastomer), the geometry of the mixer, and the operating

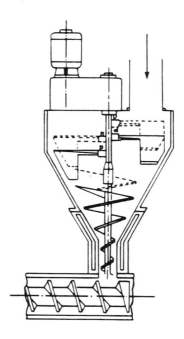

FIGURE 4.   The Sikoplast screw and hopper configuration.

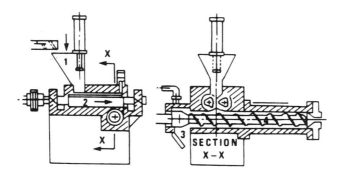

FIGURE 5.   The Farrel M.V.X. mixer.

conditions. In a specific case, any one of the three may be the efficiency-determining step. Simple mixing or homogenization involves moving a particle from one point to another, without changing its physical shape. This increase in randomness or entropy is referred to as extensive mixing. If the shear forces are sufficiently large, particles may fracture (dispersive mixing), and the polymer may flow (laminar mixing). In addition, if the deformation of the elastomer exceeds its breaking strain, then it will break into supermolecular flow units.

There are four physical changes, from the point of view of the ingredients in the mix, which take place during the mixing cycle, namely, incorporation, dispersion, distribution, and plasticization. During the initial stage of a mixing cycle, the rubber is forced or drawn into the working area between pairs of rotors and between rotors and chamber wall, and the identity of the original rubber bales or particles is destroyed. The incorporation stage occurs when the initially free ingredients become attached to the rubber. This is also known as the wetting stage and is comprised of two mechanisms. In the first, the elastomer undergoes a large deformation, increasing the surface area for accepting filler agglomerates, and then sealing them inside. In the second, the elastomer breaks down into small pieces and mixes with the filler and once again seals the filler inside.

During dispersion, the filler agglomerates are gradually broken down, distributed through the rubber (by simple mixing) and are then dispersed (i.e., broken down to the ultimate size), giving a fine scale of mixing. This is especially important in the case of carbon black because at this stage an intimate contact between the surface of the carbon black and the elastomer develops, resulting in bound rubber. Both the disruption of the filler aggregates and the forcing of the elastomer onto the filler surface require high shear stress. However, the shear stresses do not all result from the imposed shear field, because microscopic shear fields are also generated from elongational deformations.

The process of distribution involves increasing homogenization which takes place throughout the mixing cycle. Finally, in plasticization, the rheological properties are modified to suit subsequent operations.

It is important to note that overmixing can result in exposure to shearing and high temperatures. This results in excessive carbon black interaction, viscosity increases, and in some cases, reversion of the rubber.

There are three standard methods of mixing a rubber compound in an internal mixer, namely, the so-called conventional method, the rapid oil addition method, and the upside-down mix method. Many variations of these three methods are also used to suit the special characteristics of individual formulations and equipment. It is, in general, necessary to add particulate fillers early in the mixing cycle, so that good dispersion is achieved as a result of the high shear stress and high viscosity at the lower temperatures then prevailing. Also, oils and plasticizers, which reduce viscosity, should be added later. Upside-down procedures and variants of it are attempts to implement these ideas in practice.

The conventional mixing method consists of adding the elastomer first, then the dry ingredients, then the liquid ingredients after the dry materials are well dispersed in the elastomer. This method can achieve a homogenous dispersion of all ingredients, including fillers of very small particle size. Mixing times required are long, because it is more difficult to incorporate the liquid ingredients once the dry materials are dispersed. With fillers that are of low bulk density or with fillers that cake when dry, a variation of this technique is to add part of the liquid ingredients at the same time as the dry ingredients.

The rapid oil addition method involves adding the elastomer first and the dry ingredients as soon afterward as possible. After about 1 to 2 min of mixing, all fluids are added together. The proper time for addition of liquids needs to be determined. Dispersion usually improves if the addition of liquids is delayed slightly; however, this will extend the mixing time. Use of this method can give very good dispersion if liquids are added at the proper time. This method often is used for compounds containing a large volume of liquid plasticizers. It can lead to an extended mixing cycle due to the lubricating effect of the liquids between rubber and the metal parts of the mixer.

The upside-down mix method is the fastest and the simplest way of mixing. It is especially effective for those compounds containing a large volume of liquid plasticizers and large particle-size fillers. This method involves adding all ingredients to the mixer before lowering the ram and commencement of mixing. All dry ingredients are added to the mixer first, then all liquids, and finally all elastomers. This method is not suitable for those compounds containing low structure, small particle-size carbon black, or compounds having high loadings of both soft mineral filler and oil, together with an elastomer of high Mooney viscosity.

Premature vulcanization of a rubber compound, known as scorch, can occur if the reaction temperature of the vulcanizing ingredients is reached before the desired time. If this temperature is reached in the mixing process before the proper viscosity and level of dispersion is obtained, then the addition of accelerators and vulcanizing agents will cause scorchiness and poor processibility. Therefore, if this occurs, the batch must be mixed to the required viscosity and dispersion without the vulcanizing ingredients, emptied from the mixer and allowed to cool down. The batch is then fed back to the mixer for the addition of the vulcanizing ingredients in a second pass. This is known as two-pass mixing. In general, a compound requires two-pass mixing if satisfactory dispersion cannot be achieved below 120°C (250°F). Compounds containing high melting point ingredients need a high temperature to flux such components. This precludes the addition of curatives in the first pass.

The most common methods of deciding when to end the mixing process have been a preselected time, the compound reaching a preselected temperature, or the compound reaching a given temperature at a given time. The aim is to guarantee the quality of the end product, avoid overmixing, and reduce variation between batches.

Variations in power consumption in a typical rubber mix cycle are indicative of stages in the process, such as wetting, dispersion, and plasticization, and can be related to the development of end product properties. Properties such as viscosity and die swell, which are related to the volume fraction of rubber in the mix, reach their optimum at the end of the dispersion stages. At the beginning of the dispersion stage, the filler particles are still in agglomerates which contain rubber occluded between the particles. When subjected to shear, the entire filler agglomerate with its occluded rubber behaves as a single filler particle. As a result, a poorly dispersed mixture always has a higher viscosity and a lower die swell than a homogenous compound. Monitoring the power usage is equivalent to determining the black incorporation time.

Mixing to a preset time does not allow for variations in metal temperature at the start of the mix, cooling rate, or time of addition of compounding ingredients, and can result in significant batch-to-batch variation. When mixing to a predetermined temperature, as is often done with upside-down mixes with short mixing cycles, the major limitation is the accuracy with which the batch temperature can be measured. The large heat sink provided by the mixer often makes temperature measurement inaccurate. Mixing to a predetermined power input into the batch overcomes these limitations and gives improved batch-to-batch consistency with mixes requiring longer than 3 min mixing time.

There are several parameters that are important to internal mixing operations. The first of these is shear stress. In general high shear stresses and low particle-to-particle attraction increase the rate of dispersion. However, for a given particle there is a critical stress below which dispersion will not occur. The minimum shear stress occurs in the nip and is inversely proportional to the square of the gap, so this clearance is critical. The importance of elongational flow, which may be more efficient in particle size reduction, must not be ignored.

Both the shear strain and rate of stress strain also impact on the mixing process. In general, the total shear strain (or deformation) necessary for a particular degree of mixing can be imposed at any rate, the shorter time required at high rates being offset by higher power consumption and heat generation. Shear strain rates for a variety of types and sizes of mixers approximate to v/h, where v is the peripheral rotor speed and h is the rotor tip clearance. This means that small mixers must be run at higher rotational frequencies in order to achieve the same shear strain rate as larger mixers.

Another parameter is rotor speed, which directly affects total shear strain or deformation and thus the speed of mixing. Mixing speed is limited by the maximum allowable temperature, which is determined by the balance between heat generation and heat conduction or removal. Dispersive mixing, although dependent on shear stress, does not seem to be directly affected by rotor speed. This may be due to the dominance of elongational flow, which also creates high shear stress.

The purpose of the mixer ram is to keep the ingredients in the mixing area. Hence, incremented pressure would not be expected to influence mixing efficiency. In practice, however, high ram pressure has definite advantages, especially for high viscosity mixes. High ram pressures decrease voids within the mixture and increase shear stress by reducing slippage. Additionally, the effect of increasing pressure is to increase the contact force between the rubber and the rotor surface, thus increasing the critical stress so that flow begins at a lower temperature.

Perhaps the most important factor in mixing is temperature control. The rate at which the stock is heated and cooled, and the length of time that it is kept at each temperature, has a significant effect on its subsequent performance. The many variables involved include batch size, temperature and flow rate of the heating or cooling medium, rate of heat transfer, and work input to the stock. Care must be taken to ensure that the temperature control capabilities of the equipment are not exceeded. The equipment should be inspected periodically to ensure that the heat control mechanisms continue to operate efficiently. A major concern in most mixing operations has been to control the temperature at the end of the mixing cycle, therefore, normally cold or even chilled water has been used as a coolant. With the advent of internal mixers with more efficient heat exchange capability, cold water can be too efficient a coolant, and at the beginning of mixing the rubber slips on the cold metal surfaces. This problem can be avoided by maintaining the cooling water at a temperature such that the rubber adheres to the rotors and deforms readily. This results in more consistent mixing and a slight reduction in mixing time.

Milling was the earliest method used to mix rubber and is still widely used. In mill mixing, temperature control is critical. Cooling is usually accomplished by flooding or spraying the inside of the mill rolls with water, or by circulating water through channels drilled in the roll walls. The compound temperature is adjusted by regulating the rate of water flow through the rolls. Steam heating is used where a temperature increase is required. Roll temperatures suitable for mixing a given compound depend upon the nature of the rubber and such factors as the types and quantities of fillers and plasticizers to be incorporated. Lightly loaded ethylene propylene diene monomer (EPDM) compounds are usually milled with roll temperatures in the range 10 to 30°C (50 to 90°F); and butyl and highly loaded EPDM compounds at 60 to 80°C (140 to 180°F). For butyl and EPDM compounds, the front roll should be approximately 10°C (20°F) cooler than the back roll because these elastomers tend to release from the hotter roll. In mill mixing, homogenization is caused by the shearing action in the nip between the two rolls. During operation the elastomer is first added to the nip at the top of the rolls. A band of elastomer then comes through the nip and is formed preferably around the front roll. Depending on the elastomer, varying degrees of difficulty may be encountered in forming the band at the beginning. After a few passes a band will form and be fed back into the nip continuously. The elastomer is then cut back and forth twice to assure proper blending and to allow the elastomer in the bank to go through the nip. It is important for efficient mixing to maintain a rolling bank on the mill during incorporation of ingredients. All dry ingredients except the fillers and the cure system are then added to the nip and the compound is cross-cut back and forth twice to assure good dispersion of these dry ingredients throughout the batch. The next state involves opening the mill slightly and adding the fillers slowly to the batch. In order to prevent excessive loading of fillers at the center of the mill, strips of compound are cut from the end of rolls several times during this operation and thrown back into the bank. When most of the fillers have dispersed in the compound, the liquid and the remaining fillers are added slowly and alternately to the batch. When no loose filler is visible, the batch is cross-cut back and forth twice more to assure good dispersion of fillers and plasticizers. It is often a useful procedure to "pig" roll the batch and feed the "pigs" back into the nip at right angles as part of the cross-blending process. The final step is to open the mill more and add vulcanizing agents to the batch. When the vulcanizing agents are well dispersed, the entire batch is cut back and forth at least five times to assure thorough cross-blending before being sheeted off the mill.

Mill mixing is a slow process that requires constant physical effort from a mill operator. It is not only time-consuming, but difficult to control because it depends heavily upon the skill of the operator. The internal mixer is more widely used in the industry because of its versatility, rapid mixing, and large throughput.

The normal procedure for handling the product from a batch mixer is to dump it from the mixer onto a mill. The purpose of the mill is twofold: it cools the batch and by banding on the mill roll, changes its physical shape so it can be sheeted or stripped off and fed to a conveyor. The majority of mills used have fixed speed, fixed friction ratio and a manually adjustable gap setting for the nip. Also, for ease of stripping, the front roll is usually at the slower speed. The behavior of a particular compound on the sheeter mill depends on the balance between its viscous and viscoelastic properties. If the stresses caused by deformation in the nip exceed a certain critical value, then in the time of revolution of the roll (2 to 4 s normally), the material can relax and sag in a process akin to melt-fracture in extrusion. The stresses (or stored energy) may be high enough to initiate tears. The surface speed of the loaded roll determines the time the material has to sag and the time the tears have to spread. Hence, a large difference in speed between the rolls, a narrow nip and the material on the slower roll can lead to "bagging". In bagging, the material essentially is given too much time to relax. Unfortunately, the reverse can be true; if the material is transferred to the fast roll, the time to relax is shorter and the stresses do not relax enough, but build up with further passes through the nip resulting in tearing and crumbling. There are four regions of mill behavior depending on the operating temperature and the particular elastomer involved. Figure 6 shows the change from region 1 to region 4 as the temperature is increased. It is only in region 2 that stable milling and adequate mixing take place. At a given temperature, whether the

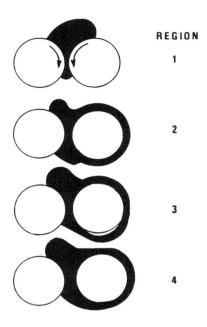

FIGURE 6.   The effect of temperature on rubber being processed on a mill.

material goes to the slow roll or the fast roll, or even drops down without banding depends on the size and speed of the mill and the rheological properties of the material. However, in general, materials tend to go to the fast roll when the nip opening is very small. When the nip opening is increased to a certain size, the material goes to the front roll. At the transition point, the material can band on neither roll, but drops down without banding. Figure 7 illustrates the effect of nip opening on the behavior of rubber on a mill.

Although continuous mixing is important, it is important to note that it is not usual practice in the rubber industry. Interest in continuous mixing has grown because a series of interconnected processes from raw rubber to finished product should lead to savings in time, labor, and energy. Semi- and fully continuous systems of mixing are only viable with long runs of a limited number of compounds. Semicontinuous mixing is the term used to describe a system in which batches from an internal mixer are discharged into an extruder. Fully continuous mixing requires the use of elastomer in powder or particulate form. A limited range of elastomers is commercially available in this form; others may be granulated or powdered in house by the processor.

Two main routes can be followed in processing powdered rubbers. These involve using heavy-duty mixers and by-passing heavy-duty mixers. The economics of processing powdered rubber are so dependent on the type of rubber, the compound formulation, the mixing equipment, the processing equipment, and the end use, that there is no general answer to the question

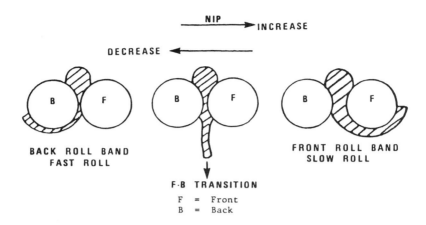

FIGURE 7. The effect of nip opening on rubber being processed on a mill.

of whether the savings obtainable in processing offset the premium normally charged for powdered rubber or crumb over the price for bale form.

III. EXTRUDERS AND THE EXTRUSION PROCESS

Extruders are basically machines that force rubber through a nozzle or die to provide a profiled material. Machines can be classified into two types: those in which the pressure required to force the rubber through the die is produced by a ram, and those in which the pressure is produced by a screw. Ram-type extruders are limited to specialized machines such as a Barwell performer. Screw-type extruders are most widely used in the rubber industry. Extruders have been in use for over 150 years. In 1845 ram extruders were invented to cover wires with gutta percha. In the 1880s, screw extruders were invented and used to process components for pneumatic tires. Other devices of lesser economic importance for polymer melt have also been developed. During the 1950s a ''normal stress pump'' was first introduced. In this apparatus melt was fed to a gap between a rotating disc and a stationary member with a die in its center.

The screw extruder is important because it continuously converts feed to a finished form, such as a rod, tube, or profile. Feed can be in the form of strip or granules. The apparatus that performs the conversion is an Archimedian screw which fits inside a barrel. Along the barrel are a number of collars that form heating/cooling zones. Feed material is forced forward by the rotating screw. As it moves forward, the feed is softened by the frictional heat developed through the shearing action of the screw, supplemented by the heat conducted from the barrel wall. By the time the feed reaches the end of the screw, it is in a viscous state that can be forced through an orifice or die and formed into the desired shape. A considerable source of power must

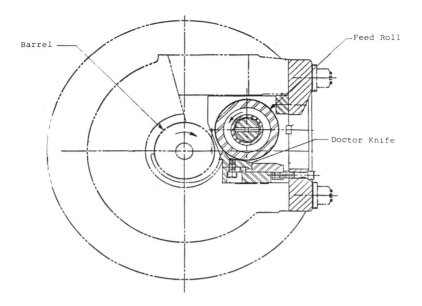

Barrel

Feed Roll

Doctor Knife

FIGURE 8. A feed roll.

be made available to drive the material forward and to develop the heat needed. For a modern 4 1/5-in. cold feed extruder, a motor of about 150 hp is required.

In the simplest form, a screw extruder consists of three components: (1) a feed screw operating within a barrel, (2) a head, and (3) a die or a die plate. An electric motor turns the screw through a reduction gear. Rubber compound enters at the feed hopper, is pushed along the barrel where it builds up pressure and temperature, and exits at the die where it takes the shape of the die. A feed hopper is provided to receive the rubber compound and guide it down into the "feed flights" of the screw. The compound may be supplied hot, in the form of intermittent or continuous strip, or it may be supplied cold in the form of a strip or pellet.

The barrel along the "feed flights" may be undercut to assist the feed. A driven feed roll positioned parallel to the screw may be provided to pull feed into the screw (see Figures 8 and 9). In some cases, the rubber compound is supplied as the discharge from an internal mixer, in which case a power-operated ram is used to push the compound into the screw flights.

If the feed volume is gradually increased, a point will be reached at which the extruder will choke and feed will "back-up" into the feed hopper. There is an upper limit to the rate of feed for a particular screw speed, and usually the best conditions for extrusion are obtained when the compound is fed at about 90% of the amount to choke the machine. The rate of feed must be determined for each compound and screw speed.

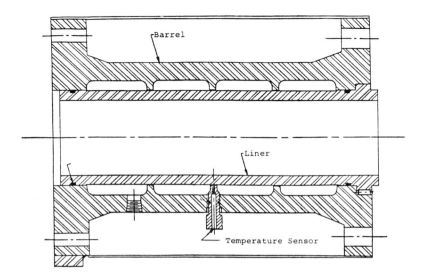

FIGURE 9. A barrel section line and temperature sensor.

The screw should preferably have a lower volume in the flights at the discharge end. This is achieved by (1) a reduction in pitch of the screw, (2) a reduction in the depth of the base of the screw, (3) a reduction in overall diameter of the screw and barrel, or (4) an increase in the number of starts in the screw. Methods (2) and (4) are the most economical and most commonly used. The screw must be full at the discharge end to avoid changes in swell of the extrudate and to maintain uniform and consistent output.

A vacuum device can be fitted in the barrel to remove volatile matter from the compound. Vacuum extruders are used to fabricate for low pressure vulcanization processes, such as open steam, hot air, molten salt and fluidized bed. The removal of entrapped air and volatile matter reduces changes of porosity that develop during vulcanization.

The head on a machine has the roles of equalizing the pressure, to transport compound to the die, and to hold the screen pack, pressure, and temperature sensors, and spider if used (see Figure 10). Rubber compound must move smoothly to the die, and ideally at equal pressures and speed. Any points at which the compound does not move are known as "dead spots". Compound can cure in these "dead spot" areas, then break away to give bits of scorched compound in the extrudate. Proper head design must satisfy the following requirements:

1. It must alter the cross-section of flow, giving it a shape corresponding to the cross-section of the extruded product.

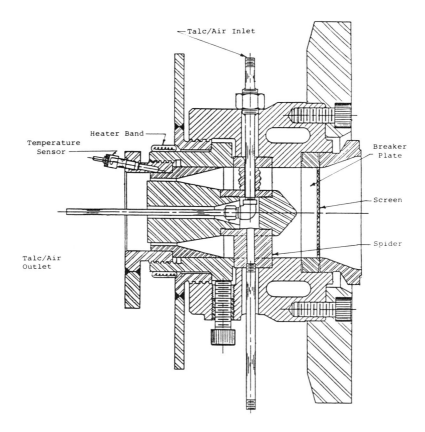

FIGURE 10. Typical extruder head with tube die.

2. Configuration of the annular gap in the extruder head which shapes the compound must take account of the alteration in shape to which the compound is subjected as a result of high elastic recovery.
3. Geometrical dimensions of the annular gap and the angle of discharge must ensure maximum output without any ''elastic turbulence''.
4. Configuration of the channels in the extruder head through which the compound flows must be such that there is no risk of forming zones of stagnation.
5. The extruder head must create sufficient resistance to develop, at the delivery end of the screw, a backpressure necessary for effective mixing and homogenization of the compound.

The die forms the compound into the desired shape of the article. Rubber compounds can exhibit significant die swell as they exit the die. This makes die design semiempirical. For each compound the amount of die swell is

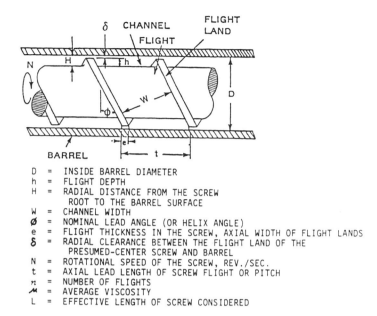

D = INSIDE BARREL DIAMETER
h = FLIGHT DEPTH
H = RADIAL DISTANCE FROM THE SCREW
      ROOT TO THE BARREL SURFACE
W = CHANNEL WIDTH
∅ = NOMINAL LEAD ANGLE (OR HELIX ANGLE)
e = FLIGHT THICKNESS IN THE SCREW, AXIAL WIDTH OF FLIGHT LANDS
δ = RADIAL CLEARANCE BETWEEN THE FLIGHT LAND OF THE
      PRESUMED-CENTER SCREW AND BARREL
N = ROTATIONAL SPEED OF THE SCREW, REV./SEC.
t = AXIAL LEAD LENGTH OF SCREW FLIGHT OR PITCH
n = NUMBER OF FLIGHTS
µ = AVERAGE VISCOSITY
L = EFFECTIVE LENGTH OF SCREW CONSIDERED

FIGURE 11. Terminology for describing typical extruder flow channel.

dependent upon the shear rate and viscosity of the compound. Die swell should be determined at production speed to duplicate shear rate and viscosity. Measurements made with a capillary rheometer or torque rheometer can be used to predict die swell at various shear rates.

For rubber processing an extruder is a single screw machine where the screw rotates in a tightly fitted barrel, thereby transmitting mechanical energy into the rubber compound. As the temperature of the rubber increases, the plasticity is lowered to such a level that the rubber can be shaped by passing through a die located at the end of the barrel. The flights on the screw and the inner surface of the barrel form a conduit or flow channel. Figure 11 shows the geometry of the flow channel and descriptive symbols for the screw and barrel.

The polymer melt transport toward the discharge end of an open-end extruder is a function of the frictional forces in the flow channel due to the rotation of the screw. In order for the material to move in the axial direction, the frictional force at the barrel must be greater than the sum of the flight and channel frictional forces. If the coefficient of friction between the material and screw is much larger than the coefficient of friction between the material and barrel, the material will turn with the screw within the stationary barrel, and the axial velocity of the elastomeric plug would be zero. In reality, the elastomeric plug moves partially in the circumferential direction and partially in the axial direction. There are three distinct sections, namely, the feed section, the transition or metering section, and the compression section. A screw for a nonvented cold feed extruder is shown in Figure 12. Each screw section performs a distinct function. The feed section transports material from the hopper. The transition section heats and mixes the material. The compression section homogenizes and builds up pressure necessary to force material through the die.

A well-designed extruder for high performance is capable of handling rubber compounds with a wide range of viscosity and processing characteristics, achieving good homogeneity regardless of extrusion speed, and providing high output rate without causing rapid rise in stock temperature. The difference in extruder design reflects different approaches toward these objectives.

With a cold feed extruder the polymer is generally fed at room temperature. The feed can be in the form of strips or pellets. The screw must transmit sufficient mechanical energy to plasticize the compound to near minimum viscosity as well as to overcome the head restriction. The cold feed extruder receives cold feed at near-maximum viscosity and reduces the viscosity to the equivalent of hot feed within a 30- to 120-s time frame.

Cold feed extruder screws require special design considerations. To accomplish the extra mastication, the flight depth (h) must be small and screw length (L) must be long. A large drive motor and gear box are required to supply and transmit the mechanical energy to the screw. The shallower flight depth (h), the higher starting viscosity of the compound, and the longer screw (L) have a combined effect that makes the cold feed extruder less sensitive to flow variations with changes in headpressure (P). Plastication is accomplished in the transition section. Reduction in the flight depth (h), high compression ratios, and increase in the number of starts are the variables used to achieve high shear and high output. Figure 12 shows a typical screw for nonvented cold feed extruders.

In general, an extruder that is fed rubber compound at a temperature above ambient is classified as a hot feed extruder. A hot feed extruder receives rubber compound at near-minimum viscosity, and must overcome the head restriction with a minimum temperature rise and hold time. This is necessary because the compound has already been at an elevated temperature for a period of time during milling.

The overall objective in hot feed extruder designs is to minimize the temperature rise of the compound through the extruder. The flight depth of the feed screw must be fairly high to hold down the energy imparted to the compound and to keep its temperature low. Unfortunately, as headpressure increases, output decreases. Therefore, hot feed machines are most suitable only for applications requiring low headpressure. Figure 13 shows a typical screw for hot feed extruder.

The vented cold feed extruder (also called a vacuum cold feed extruder) was developed to expel unwanted atmospheric pressure. The screw in this

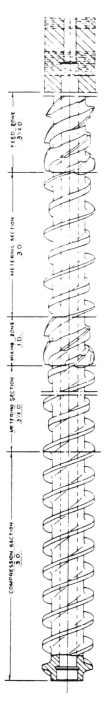

FIGURE 12.   A screw for nonvented cold extruder.

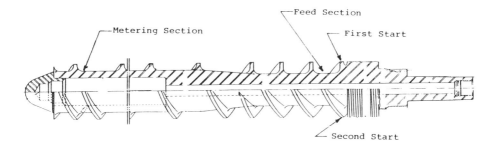

FIGURE 13. A typical screw for a hot feed extruder.

design has two distinct zones. The first zone has three sections: feed, transition, and metering. The second zone has two sections: transition and metering. A dam separates the two sections. Figure 14 shows that each of these sections performs a distinct function. The feed section moves material from the hopper or feed box into the barrel. Heating and fluxing occur in the transition section. Homogenizing and build-up of pressure occur in the metering section. In the dam section, small axial grooves cut in the dam surface serve as dies. A vacuum pump connected to a port located below the exit side of the dam maintains a low pressure zone. Material exits the dam zone in the form of many thin strands and enters the second transition section. The combination of thin strands, high temperature, and low pressure are conducive to the release and escape of trapped gases. Escape of trapped gases and further heating and fluxing occurs in the second transition section. Homogenizing and buildup of pressure necessary to force material through a die is accomplished in the second metering section. Plugging the vent port with compound can occur when headpressure rises to a level at which the first zone pumps at a higher rate than the second zone. Temperature profiling can be of some assistance to reduce the pumping rate in the first zone and increase the pumping rate in the second zone. Increasing the L/D ratio in the second zone is an effective way to increase the pumping rate in the second zone.

The pin extruder is another variation in extruder design. Rotating laminar planes are generated around the flow channel of a simple conveying screw and little exchange of material occurs between these layers. The warm layers of material that contact the screw and barrel surfaces remain stratified. The cold rubber at the core is slow to warm up because it is insulated from the warm outer layer due to the low thermal conductivity of rubber. This effect exists along the entire length of the screw, causing difficulties in attaining optimum homogenization. An uneven temperature profile develops in the cross-section of the extrudate as corrugated surfaces as well as unacceptable dimensional variation. This effect grows as the flight depth, pitch angle, and diameter of a conventional screw with constant pitch and root diameter increases. The larger the depth of the screw channel the more difficult it becomes

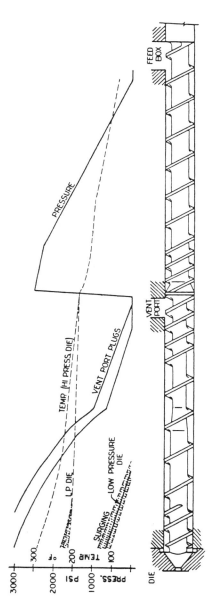

FIGURE 14. A typical screw for a vented cold feed extruder.

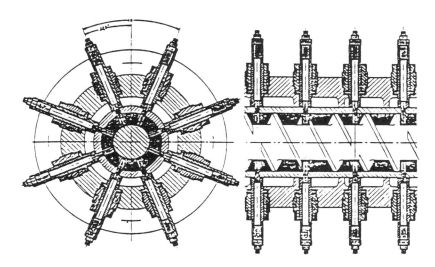

FIGURE 15. Cross-sections of pins in both radial and axial arrangements.

to uniformly homogenize the compound through friction from the barrel and screw surfaces. By mounting pins in the barrel it is possible to provide an effective way to interrupt or split the laminar planes and improve homogenization without using high shear rates. Slipping on the barrel wall is also minimized by the pins; 8 to 12 pins are arranged radially in 6 to 10 rows (see Figure 15). The first pin row is located near the feed inlet where the screw filling sequence has been completed. These pins hold the cold rubber compound radially so that the conveying lands convey the compound axially. A mixing zone such as shown in Figure 16 is positioned immediately downstream of the last row of pins. This mixing zone results in further homogenization.

To remove frictional heat through the screw and barrel it is necessary to transfer the heat from the rubber to the metal barrel. To achieve a good flow of heat, new surfaces of the rubber must be exposed to the metal surfaces. This is a turbulent flow condition. To achieve turbulence in the rubber, it is necessary for the rubber to grip the metal surfaces, enabling shear to take place. If rubber slides on the steel surface, more frictional heat is generated at the surfaces of heat transfer. The grip of the rubber depends on the roughness of the surface, the lubrication of the surfaces and the temperature of the rubber and metal. The best temperature is the lowest possible temperature which will result in a sufficiently high coefficient of friction. Both hot and cold feed extruders are designed to optimize temperature control. In general, cold feed extruders require more sophisticated temperature control systems. The range of temperatures that may be individually controlled by the temperature control system at different parts of the extruder is referred to as the temperature profile. In cold feed extruders, the zones separately controlled are usually the

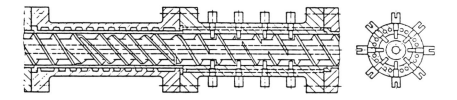

FIGURE 16. Cross-section of pins and mixing zone.

die, head, one or more barrel zones, feed zone, and the screw. In the vented extruders there is usually an additional zone of control, the vacuum zone. The frictional heat introduced by the screw is usually so great that a large proportion of the energy supplied by the motor must be removed by cooling the screw and barrel. Zone temperatures are not the same as the temperature of the rubber compound passing through that particular zone. This is due to the comparatively small amount of surface area, relative to volume, exposed in each zone and to the elapsed time the material takes to pass through the zone. Temperature control is accomplished by the use of thermocouples to sense the temperature and an electronic temperature controller to control the amount of heating and cooling.

An example of a temperature profile is shown in Figure 17 for an EPDM compound using a standard screw. It is necessary to actually process each compound in the extruder to determine the effect of temperature on its output rate, uniformity of flow, quality of extrudate, and extrudate temperature.

The extrusion process affects the output rate, extrudate temperature, Mooney viscosity, and scorch time of the compound. Maximum output is limited by the maximum permissible heat history of the compound (scorch time). At low screw speeds, output rate is nearly directly proportional to screw speed (RPM) for most compounds. As the screw speed increases toward the high end, the output rate decreases. Output rate is normally given in terms of pounds per hour or pounds per revolutions per minute.

Extrudate temperature rises as screw speed increases. The actual temperature rise is affected by screw design, head pressure, and the properties of the compound being processed. The output rate for the nonvented extruder is about double that for the vented extruder. The rise of extrudate temperature is higher at the vented extruder. Complex manufacturing processes are seldom trouble-free. Extrusion is no exception. Among the common processing problems are those relating to output rate, dimensional stability, heat generation, and rough extrudate.

Dimensional stability of the extrudates depends on several factors. First, the strength of the rubber extrudate is relatively low as it exits the extruder. Take-away equipment that includes take-off conveyors, water baths, take-up reels, etc., must be properly designed. Drive motors should be variable-speed with fine speed adjustment. Stress applied to the extrudate must be constant.

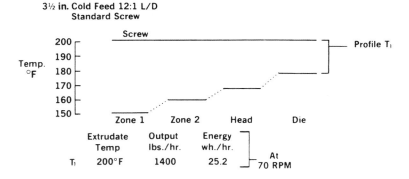

FIGURE 17. Typical temperature profile for rubber extrusion.

Cooling should be sufficient to lower the temperature of the extrudate and allow it to develop normal "green strength".

As noted above, output is roughly directly proportional to screw speed. If screw speed fluctuates or drifts, the dimensions of the extrudate will change because output is not constant. Speed of the main drive motor should vary <1%. D.C. drive systems should be set up to give this level of speed control. Electric power supplied to the drive motor should be free of surges. A speed meter should be mounted on the control panel to monitor speed accuracy.

Output will change if the feed to the extruder fluctuates. A uniform feed strip coupled with a power-driven feed roll and temperature control on the feed zone enhance the uniformity of output. A slightly flooded feed hopper is better than a starved feed hopper. The size and number of feed strips should be adjusted to maintain a small ball in the feed throat. Additionally, output will fluctuate if the temperature control system permits variations in the controlled zones. Each temperature controller should be tuned to maintain a temperature uniformity of $\pm 1°F$.

Finally, throughput and die swell are strongly dependent upon compound design and uniformity. If output or die swell changes immediately after starting a new skid of feedstock, a change in compound or batch-to-batch uniformity is suspect. Mooney scorch should be determined for both the feedstock and the extrudate.

Temperature buildup in the extrudate can result in scorchiness. Scorch is affected by extrusion speed, screw configuration, headpressure, and compound formulation. Reducing screw speed, adjusting temperature profile, using a less-mastication screw, and reducing restriction at the die are positive directions. Modifying the compound to reduce viscosity, adding prevulcanization inhibitors, and changing the cure system are other alternatives.

Finally, surface roughness of the extrudate not caused by a scorchy compound is either due to air entrapment or insufficient mastication. Both causes are related to screw design and screw speed. Adjusting the temperature profile

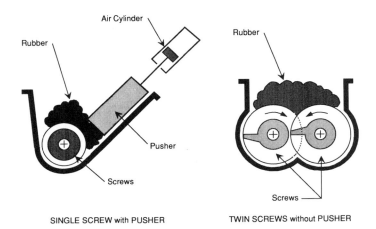

FIGURE 18. Compares single screw configuration (left) with pusher against twin screw without pusher mechanism (right). (Courtesy of Kobelco Stewart Bolling, Inc., 1600 Terex Road, Hudson, OH 44236. With permission.)

to increase mastication, changing to a screw design for more mastication, and additional breakdown of the compound in the mixer are possible solutions to this problem. In some cases polishing the inner surface of the die or increasing the length of the die land will reduce surface roughness.

The twin screw extruder has become popular in recent years, particularly because of the need for more complex profiles as in automotive seals and sponges where two different rubbers are often mated in a final profile part. Although there are several design configurations used, with advantages distinct to the particular fabricator's objective, only one is highlighted here as an example. Kobe Steel, Ltd. (Chiyoda-ku, Tokyo, Japan) has developed a twin screw-type roller head extruder. The roller head extruder is typically installed under a batch type or continuous type mixer and kneader to form a mixed rubber sheet automatically. This particular design is not equipped with any pusher which is often used in more conventional single screw extruders. This enhances the operational performance of the equipment from the viewpoints of energy saving, cooling effect, and maintenance. This extruder has a short pressure zone at the cylinder of the extruder which means that there is less heat generation. It is therefore possible to decrease the temperature of the rubber against the rubber temperature at the extruder after the calender roll. As such, the extruder can be applied widely for mastication, masterbatch, and final mixing.

For comparison, Figure 18 illustrates a single screw extruder with pusher mechanism and the twin screw configuration without pusher. The pusherless extruder provides several distinct advantages. First, as already mentioned, the energy savings can be significant, as the pusher operation compressed air is not necessary. Second, there is no sliding friction between the pusher and

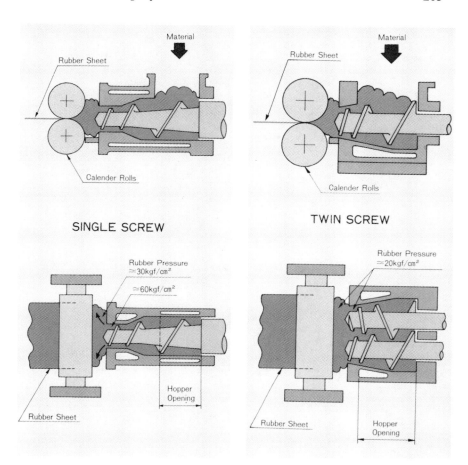

FIGURE 19. Side view comparison of single (left) and twin screw (right) system for sheet forming operations. (Courtesy of Kobelco Stewart Bolling, Inc., 1600 Terex Rd., Hudson, OH 44236. With permission.)

pusher guide. Finally, there are no rubber remnants at the pusher sliding clearance and at the top of the pusher, thus eliminating scrap.

By horizontal arrangement of the twin screws, the material outlet is widened and hence the rubber can flow smoothly at both ends of the calender roll. Since the extruded rubber can flow smoothly, a better rubber sheet can be obtained (the manufacturer claims this is possible even in the rubber pressure if the cylinder is low). Localized heating on the rubber material due to rubber pressure in the cylinder is also kept at a low level, which is beneficial in minimizing scorch. Figure 19 provides a comparison of the horizontal (side view) arrangement of single screw and twin screw in sheet forming operations.

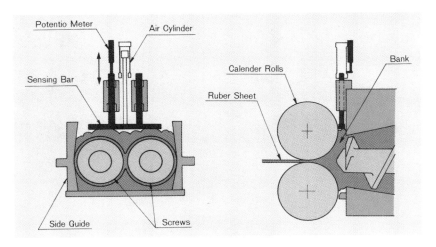

FIGURE 20. Bank control on the twin screw design. (Courtesy of Kobelco Stewart Bolling, Inc., 1600 Terex Road, Hudson, OH 44236. With permission.)

It is important to note that proper spacing (bank) is required between the outlet of the extruder and calender roll, which serves to detect the level of rubber retained at this position by a sensor which is used to provide feedback control of the calender roll rotating speed or screw speed. During normal operation the rubber level is kept constant which helps to maintain a uniform rubber sheet (i.e., thickness and width control). Figure 20 illustrates how bank control is maintained.

In general, sheet line extruders are complex equipment and commercial units demand sophisticated control/automation in order to produce a high-quality product. Figure 21 provides an example of a commercial scale unit designed for a multilayer sheet line.

IV. CALENDERING OPERATIONS

Rubber calenders consist of two or more hardened and accurately machined metal rolls rotating in bearing journal boxes which are set in rugged iron frames. At least one roll is equipped with screwdowns to control the thickness of the processed material. Adjacent pairs of rolls rotating in opposite directions form a "nip", in which the material being processed is squeezed into sheets or is laminated to form the desired product. The drives for the rolls include constant or variable-speed motors and reduction gearing to achieve roll surface speeds required by the processing requirements of the materials. A calender, depending on the number and the design of its rolls, is capable of sheeting, frictioning, coating, profiling, and embossing. A variety of roll configurations, both horizontally and vertically, are available in sizes ranging from laboratory units to commercial designs weighing tons. However, the

FIGURE 21. Multilayer sheet line system. (Courtesy of Davis-Standard, Division of Crompton & Knowles Corp. #1 Extrusion Drive, Pawcatuck, CT 06379. With permission.)

three-roll vertical calender with 24-in. (diameter) x 68-in. (face length) rolls and the 4-roll "Z" and "L" with 28-in. × 78-in. rolls are typical of the machines used for mass production of tires, belting, sheeting, and similar articles. Three-roll calenders have been widely used for processing of mechanical rubber goods. Four-roll calenders are popular in tire plants. The four rolls permit simultaneous application of rubber compound on both sides of tire cord fabrics. Two-roll calenders are used to produce strips and profiles, often in combination with extruder feeding, in which case they are commonly referred to as "roller dies".

Calenders are used in five separate operations in the manufacture of rubber products: sheeting, frictioning, coating, profiling, and embossing. Sheeting employs a two-roll calender in a horizontal or vertical configuration. The feed material, either in strip or "pig" form, is fed into one side of the nip and is flattened. The material emerges as a sheet that is pulled from the roll by some manual or mechanical means. Figure 22 illustrates basic calendering operations. Thickness control is accomplished by use of the screwdowns and may be further refined by automatic control systems using thickness sensors. The force required in the nip to flatten the feed material causes slight deflection of the rolls. If some corrective steps are not taken, the product thickness will vary across the sheet, resulting in excessive variations of the product and possibly the production of scrap.

In order to minimize air entrapment and blistering, the thickness of each sheet is generally limited. To build up the required thickness of the final

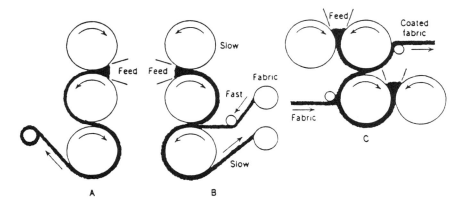

FIGURE 22.   Basic calendering operations.

sheet, two or more piles of calendered sheet are usually laminated on the bottom roll of a three-roll calender.

The operation of frictioning involves rubbing or wiping an elastomeric compound into a substrate of textile or metallic cords which may or may not be held together by "pick" threads or fill yarns, or the substrate may consist of a "square woven" fabric like "hose ducks" or "belt ducks". Usually a three-roll calender is used. The rubber sheet is formed between the upper and middle rolls while the resulting sheet is simultaneously being frictioned into the substrate between the middle and bottom rolls. In this operation the upper and middle rolls may be moving at "odd" or "even" surface speed, but the middle and bottom rolls will be run at "odd" or unequal surface speeds so that the rubber is effectively wiped into the substrate being carried on the bottom roll.

A coating or skim-coating operation is similar to that described for frictioning except that the middle and bottom rolls are operated at "even" surface speed so that the rubber sheet is laid and pressed against the substrate. In a multipass operation the substrate will have been previously frictioned. The coating operation may produce a heavy deposit or a thin "skim" coat depending upon the product requirement. Generally, multipurpose calenders such as a three-roll unit are equipped with "even" and "odd" gearing arrangements so that a number of combinations on roll speed ratios are possible. A more complex form of coating calender is the four-roll "Z" or "L" arrangement. A four-roll calender can simultaneously apply a rubber coating onto both sides of a fabric. In effect, the no. 1 and 2 rolls and the no. 3 and 4 rolls form pairs from which two rubber sheets are produced. The sheets are then laminated to a substrate between the no. 2 and 3 rolls.

Many rubber products require uncured components that are not rectangular in cross-section. In such cases, at least one roll may have a peripheral design cut into its surface to produce the desired cross-section. This method is useful when long production runs are possible, but becomes expensive in terms of roll change and roll inventory necessary when many different sections are required. In this instance, the calender roll may consist of a heavy basic mandrel onto which may be clamped solid cylindrical or split cylindrical steel "shells" into which the appropriate profile design has been cut. This operation is known as profiling.

Some rubber products are made from uncured components which must have a surface design that cannot be economically formed by subsequent molding. One such example is the cover strip around the sole of canvas shoes. The method of producing such strips is similar to that of profiling in that the required design is engraved into the calender roll or "shell" as a mirror image of the design itself. This operation is known as embossing.

Calenders are commercially available with rolls in various diameters, face lengths, and configurations plus devices to control process temperature and product dimensions. Figure 23 shows possible arrangements of calender rolls as well as standard sizes available in the rubber industry. Arrangement of the rolls is generally determined by: (1) the resolution of separating forces between rolls and roll deflection, (2) ease of rubber stock feeding, (3) space requirements for calender-mounted attachments such as trim knives and gauging sensors, and (4) space requirements for auxiliary attachments.

As already noted, calender designs differ from one another by the arrangement of their rolls and depend on the materials to be processed. For example, inverted L-type calenders are predominantly used for the production of soft polyvinyl chloride (PVC) films. In contrast, if we were making rigid PVC films, this operation would best be performed on L-type calenders equipped with four or five rolls.

Roll dimensions (i.e., the ratio "roll diameter-roll face length") are established by the roll separating forces, which in turn are dictated by the material to be processed, as well as by the thickness and width of the film. Computers are used to evaluate these values from a multitude of parameters, and subsequently the roll dimensions are determined in accordance with the roll material. Rolls are generally commercially available in incremental diameters (typically in 50-mm steps). The roll face length is calculated on the basis of the required finished width of the sheet or film to be produced.

Figures 24 and 25 illustrate standard calender lines for the manufacture of PVC films. In these designs a planetary gear extruder is used with a subsequently arranged mill. In these examples, the systems are used for the processing of a premixed PVC dry blend. The PVC compound is carefully plasticized in a planetary gear extruder and then homogenized and degassed in the mill. The mills primary functions are, however, the preparation of reclaimed material (cold or warm edge trimmings, start-up, and/or other rejects) to be refed into the calender, and to function as a material buffer zone when changing the recipe.

CALENDERS — STANDARD SIZES

Roll Diameter	Conventional Normal Journal Diameter	Conventional Normal Face
10	6.5	20
12	8	24
14	9.25	30
16	10.5	36
18	12	48
20	13.25	54
22	14.5	60
24	16	68
28	19	78
30	20	84
32	21.5	92
36	24	96

Inches used in all dimensions.

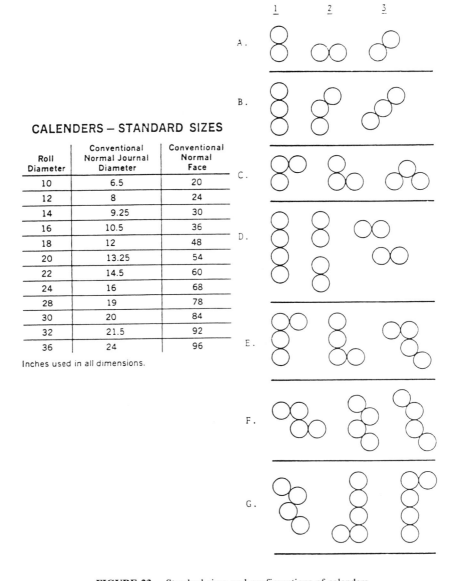

FIGURE 23.   Standard sizes and configurations of calenders.

The material strip from the mill is conveyed via a metal detector into the strainer where it continues to be homogenized and screened from contamination. The first roll gap of the four-roll inverted L-type calender is fed by a wigwag conveyer belt. The final form of the film is obtained in roll gaps 2 and 3. The film is then taken off the last calender roll by the

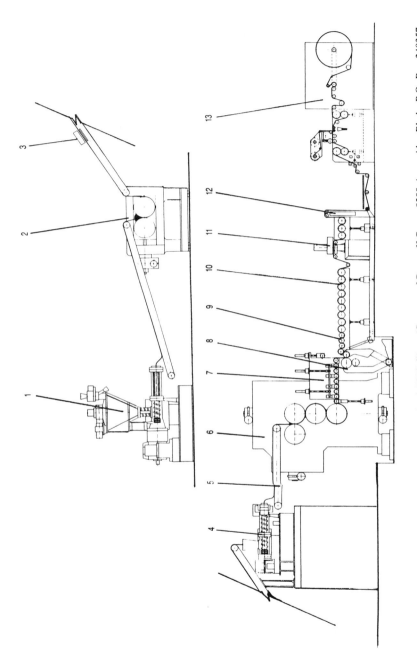

FIGURE 24. A calender line for the manufacture of soft PVC films. (Courtesy of Berstorff Corp., 8200 Arrowridge Blvd., P.O. Box 240357, Charlotte, NC 28224. With permission.)

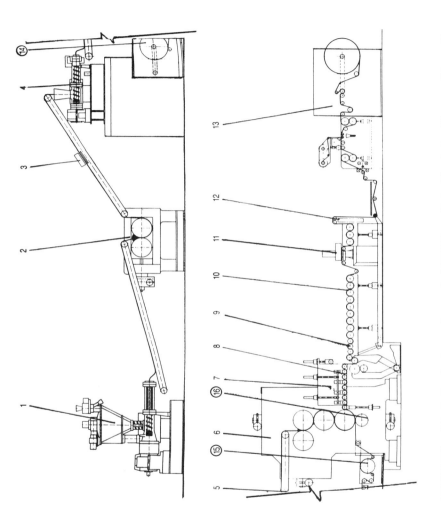

FIGURE 25. Calender line for the manufacture of soft PVC films with doubling device. (Courtesy of Berstorff Corp., 8200 Arrowridge Blvd., P.O. Box 240357, Charlotte, NC 28224. With permission.)

multiroll pick-off device. The subsequently arranged embossing unit provides the film with the requested surface structure. The take-off device takes the film from the embossing unit and leads it into the heating and cooling unit to be gradually cooled down in a controlled manner. A Beta gauge is arranged at the end of the cooling unit. Here, all film profile data are recorded and transmitted to the process computer automatically controlling the last calender roll gap, the pick-off speed, the axis crossing, and the roll bending device.

A luminous screen is provided for the final visual inspection of the film before the latter is wound up by the automatic double turret-type winder. The winder is process controlled as well so that all functions are carried out automatically.

The distinction between Figures 24 and 25 is that the latter depicts a doubling unit that enables doubling of the film either with fabric, nonwoven, or another film. The numbers in each of these schematics identify the following components:

1. Planetary gear extruder
2. Mill with mixing and turning device
3. Metal detector
4. Strainer
5. Pivot-mounted feeding device
6. Four-roll inverted L-type calender
7. Multiroll pick-off device
8. Embossing unit
9. Take-off device
10. Heating and cooling unit
11. Beta gauge
12. Inspection screen
13. Double turret-type winding machine
14. Double film let-off device
15. Preheating drum
16. Doubling unit

Calender lines for the manufacture of rigid PVC films differ from those for soft PVC films in their design and additional equipment, depending on which production process is chosen. These systems are illustrated in Figures 26 and 27. The metal detector in front of the calender detects any metal particles in the material. A separating unit cuts out the metal-containing section of the material and removes it. The calender — in this case equipped with four rolls in L-type arrangement — forms the compound to a film. Depending on the process-technical task, calenders with five, six, or in special cases, seven rolls can be used.

The multiroll pick-off device takes the film and leads it to the embossing unit where it is provided with the requested surface structure. The subsequently

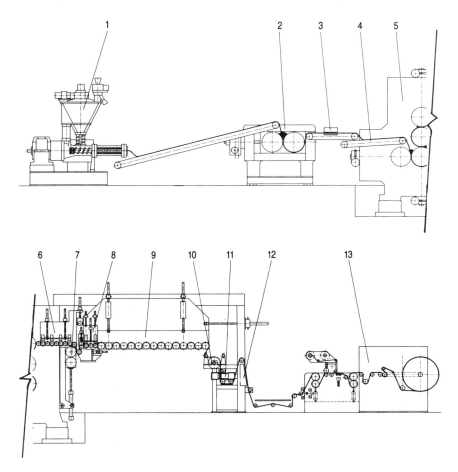

FIGURE 26. Calendering line for the manufacture of rigid PVC films. (Courtesy of Berstorff Corp., 8200 Arrowridge Blvd., P.O. Box 240357, Charlotte, NC 28224. With permission.)

arranged equipment is the take-off device, the heating and cooling unit, the Beta gauge, which in conjunction with a process computer controls the calender line, and finally the processor-controlled automatic double turret-type winding machine.

In Figures 26 and 27, the numbering scheme identifies the following equipment:

1. Extruder
2. Mill with mixing and turning device
3. Metal detector
4. Pivot-mounted feeding device

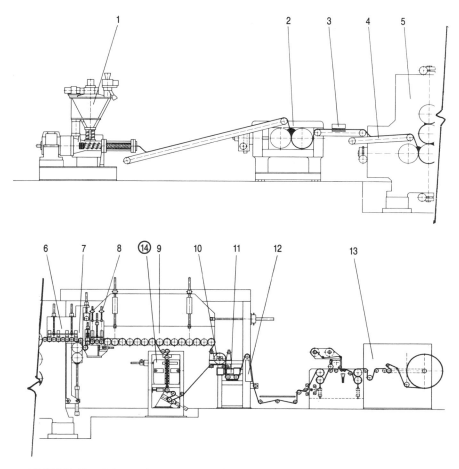

FIGURE 27. Calender line for the manufacture of rigid PVC films additionally with roller stretching unit. (Courtesy of Berstorff Corp., 8200 Arrowridge Blvd., P.O. Box 240357, Charlotte, NC 28224. With permission.)

5. Four-roll L-type calender
6. Multiroll pick-off device
7. Embossing unit
8. Take-off device
9. Heating and cooling unit
10. Vacuum roll
11. Beta gauge
12. Inspection screen
13. Double turret-type winding machine
14. Roller stretching unit

High-grade composites and films from a multitude of thermoplastics and synthetic rubber compounds can be manufactured with laminating and coating units that are interfaced with calenders. Figure 28 illustrates a combined laminating and double line (laminating) arrangement. In this figure the numbers correspond to the following equipment:

1. Double let-off devices with wind-up devices for protective cloth
2. Pull-roll and hold-back stand
3. Sewing machine
4. Dance for the control of the material tension
5. Material accumulator
6. Tension measuring unit
7. Cascade extruder
8. Conveying unit
9. Preheating roll
10. Heating screens
11. Three-roll calender
12. Laminating and embossing unit
13. Tension measuring device
14. Cooling unit I
15. Beta gauge
16. Trimming device
17. Cooling unit II combined with hold-back stand
18. Dancer for the control of the material tension
19. Material accumulator
20. Pull-roll and hold-back stand
21. Automatic double turret-type winding machine

While the extruders in cascade are rated for an optimum preparation of the PVC compound, the calender itself prepares only the sheeting leading to higher output capacities and considerable quality improvements of the products. In front of the calender, a complete fabric section is provided with a double let-off device, a gluing and sewing station, a material accumulator with the relevant pull-roll and hold-back stands, an edge control unit, and a fabric preheating device. The train behind the calender consists of the laminating and embossing unit, cooling unit, Beta gauge, trimming device, material accumulator, pull-roll and hold-back stand, and the automatic double turret-type winding machine. The same line can also be used as a doubling unit only, as shown in the diagram. The doubling is then effected by means of the bottom calender roll and the doubling roll.

Rolls have to meet highest requirements as the quality of the film is dependent to a large extent on the roll quality. Important criteria are

• Highest accuracy of concentricity and shape, at ambient and operating temperatures

- Surface finish of high quality and uniform hardness
- Resistance against deflection
- Pressure resistance against heating agents

Depending on the loads and the design, the following rolls are typically used:

1. Processing of soft and semirigid PVC: Chilled rolls made of spheroidal graphite cast iron. Surface hardness between 440 and 480 Vickers.
2. Soft and rigid PVC: Double pured rolls with a core made of spheroidal graphite cast iron and a shell made of chilled cast iron. Surface hardness between 520 and 560 Vickers.
3. Rigid PVC (S-PVC and E-PVC with a high K-value): Steel rolls of forged and surface-hardened design. Surface hardness up to 585 Vickers. The quality of the roll surface is dictated by the application requested:

   - High polished — with a minimum surface finish of Ra = 0.04 . . . 0.06 µm for films which are embossed or matted after the calendering process
   - Superfinished — with a minimum surface finish of Ra = 0.01 . . . 0.025 µm for films with nonstructured surfaces
   - Chromium-plated and superfinished — with a minimum surface finish of Ra = 0.01 µm for crystal-clear rigid films

The rolls can be ground in hot condition, i.e., at operating temperature, in order to attain the highest accuracy of concentricity and profile during calendering.

Previously to the grinding, the deflection lines of every roll must be exactly calculated. Due to the load ratios in the calender during production, the roll deflections are determined by means of a computer program specially developed by the manufacturer. The camber required for each individual roll is determined on the basis of the integrated axis crossing and roll bending device. The operating width and the temperature curve are important parameters that also have to be considered. The curves resulting from the above calculations are guidelines for the roll grinding. On a special test stand, all finish-machined calender rolls are subjected to quality control at operating temperature. These test conditions correspond to future operating conditions. Ultramodern testing instruments determine the tolerances relevant to shape and position, and measure the surface finish as well.

The pick-off device takes the film from the last calender roll and leads it to the embossing and/or heating/cooling unit. As within the pick-off device, the temperature and speed must be precisely adapted to the film recipe, the pick-off rolls have been subdivided into several heating circuits and driving groups.

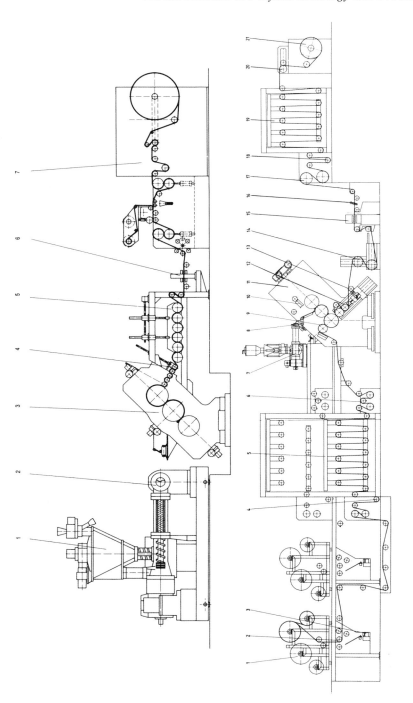

FIGURE 28. Combined laminating and doubling line (laminating). (Courtesy of Berstorff Corp., 8200 Arrowridge Blvd., P.O. Box 240357, Charlotte, NC 28224. With permission.)

The pick-off rolls are provided with a forced circulation system of the heating and/or cooling agent (hot water or oil) so as to attain a uniform roll surface temperature.

The pick-off device is driven by D.C. motors to be controlled independently from one another. The universal joint shafts and heating hoses for the rolls are led through slots machined on both sides in the calender frame. In the area outside the calender frames, the drive and heating hoses are located on one side only of the pick-off rolls.

The surface quality of the pick-off rolls depends on the process-technical requirements. Should the film be provided with a special structure, the first pick-off roll will be roughened by sandblasting or embossing and subsequently chromium plated. Embossing units are applied for rigid and soft PVC films.

For embossing units with horizontal film inlet, illustrations 1 and 2 in Figure 29 show an embossing unit whose rubberized counter roll can be exchanged against an embossing roll. Thus, it is possible to emboss either the film's bottom or top. Illustration 3 (Figure 29) shows a double embossing unit. By actuating a turnstile, a second embossing roll can be moved, during operation, into working position.

For embossing units with vertical film inlet, refer to illustrations 5 and 6 (Figure 29). The drive of these embossing units is effected to the cooled roll by means of a D.C. motor from where the driving torque is then transmitted to the rubberized and embossing rolls.

A special feature is that the diameters of the rubberized and the embossing rolls can be changed without having to correct the speed. The adjustable inlet roller ensures a very short run of the film into the embossing roll gap. In both cases, the embossing rolls can be changed during operation. Also, in the case of the embossing unit shown in illustration 6 (Figure 29), the rubberized roll can be changed during production.

Illustration 7 (Figure 29) shows the embossing unit in an open position. The roller is led through the open embossing roll gap and loosens the film from the rubberized and the embossing roll. Here also, the embossing rolls can be changed at running film production. For this procedure, an auxiliary drive is started.

The embossing devices shown in illustration 5 and 6 are mostly used for L-type calenders, while the embossing unit in illustration 4 (due to its space-saving design) is mainly used for inverted L-type calenders.

The function of the take-off device is to take the film from the embossing roll within the shortest of distances, it is almost tensionless. The device usually consists of four rolls and the design is similar to that of the pick-off rolls. The rolls are heated/cooled and driven in commonly by one D.C. motor. Generally, the take-off device is mounted to the heating/cooling unit and can be moved with it horizontally. Roll 1 can be separately adjusted in height. Thus, the take-off device can be moved toward the top edge of the embossing roll, regardless of the latter's diameter, and the film can be retrieved in a

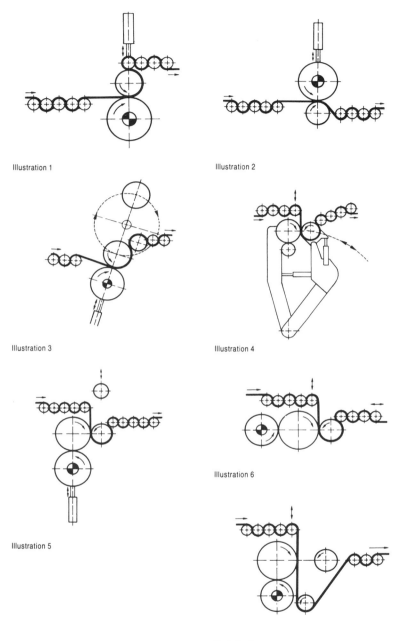

Illustration 1

Illustration 2

Illustration 3

Illustration 4

Illustration 5

Illustration 6

Illustration 7

FIGURE 29.   Embossing unit configurations. (Courtesy of Berstorff Corp., 8200 Arrowridge Blvd., P.O. Box 240357, Charlotte, NC 28224. With permission.)

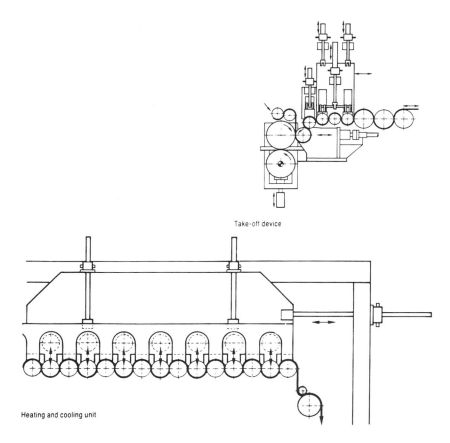

Take-off device

Heating and cooling unit

FIGURE 30. Take-off device. (Courtesy of Berstorff Corp., 8200 Arrowridge Blvd., P.O. Box 240357, Charlotte, NC 28224. With permission.)

linear and tension-free manner. If the embossing unit is not used, the complete take-off device can be lifted and moved horizontally to the last pick-off roll so as to form one unit with the pick-off device. Also in this case, a tension-free guide of the film is ensured (Figure 30).

Beta gauges operating according to the transmission measuring principle are used for the continuous control of the calendered film. Beta gauges control the film in a longitudinal direction and measure its profile in a transverse direction. The data obtained are evaluated by a process computer and indicated on a monitor.

Depending on the deviation from the set values, pulses are transmitted to the roll-adjusting or roll-bending device. The axis crossing device and the speed setting of the take-off device can also be included in the automatic control. When using large-scale computers, complete film programs can be stored.

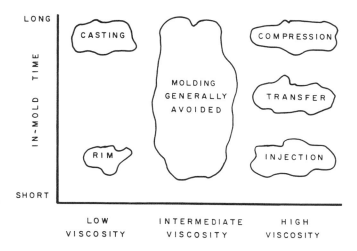

FIGURE 31.   In-mold time as a function of viscosity and molding method.

V. MOLDING

Molding is the act or process of shaping in or on a mold, or anything cast in a mold. Rubber is cross-linked during the molding process. Cross-linking is the act of connecting together the extremely long rubber molecules in a rubber composition and is usually done at high temperatures. Cross-linked molded parts are successfully used in a wide variety of consumer, industrial, and engineering applications. To meet the different needs of these applications, many different types of rubber have been developed. Different rubbers and the materials used to form parts vary considerably in their characteristics. Some materials used to mold polyurethane compositions might have very low viscosity. Most rubber used for molding has a much higher viscosity. Low viscosity materials become rubber-like only after they react with one another in a mold. In contrast, high viscosity materials, like natural rubber, are rubber-like before they enter a mold. Low and high viscosity materials are similar because they both cross-link within a mold.

Figure 31 provides a general overview of the relationship among in-mold time, molding method, and viscosity. This chart does not take into account part size which significantly affects in-mold time. Materials having intermediate viscosity (like a caulking compound) are generally avoided in molding because they are difficult to handle and show a strong tendency to trap air during processing and molding. Entrapped air is often evident as porosity in a molded part.

When a low viscosity material is cast and allowed to react partially, thus increasing its viscosity, the higher viscosity material can be molded by high pressure methods such as compression or transfer. High molding pressure acts on this high viscosity material, squeezing out the air and minimizing porosity. Thus, an article initially molded by casting might be molded finally by compression.

Compression molding is usually performed between rigid metal plates. Bladder molding is a variation of compression molding in which the material to be molded is squeezed against a rigid mold surface by a flexible rubber bladder. Bladder molding is applied in the production of tires.

During molding, rubber must flow or deform in order to fill the cavity in which it is molded. One of the ways it deforms is in shear. During injection molding, rubber often flows through a channel in its path to a mold. Shear flow in a channel can be compared to extending elements of a telescope. Instead of flat cards sliding upon one another, concentric tubes slide past one another on their common axis. Flow in a mold is complex and the mode of flow can vary. For example, extensional or stretching flow might occur in tapered sections. If water flows in a channel at sufficiently low shear rates, a ''particle'' of water moves in a straight line path; this behavior represents laminar or viscous flow. If the shear rate is increased to some critical value, however, flow becomes unstable and a ''particle'' of water moves in a turbulent manner. In the range of shear rates associated with laminar flow, water has a constant viscosity value (Newtonian behavior). For a Newtonian fluid-like water in the laminar range, the rate of flow is directly proportional to the applied pressure. Therefore, doubling the pressure doubles the flow rate. In contrast the viscosity of a typical high viscosity rubber is about 10 million times higher than that of water; thus water and rubber cannot reasonably share a common viscosity axis on a linear basis. Rubber viscosity responds quite differently to a change in shear rate. Rubber viscosity decreases sharply with increasing shear rate, i.e., it is non-Newtonian. Because of this relationship, increased pressure on rubber disproportionately increases its flow rate relative to water. This happens because higher shear rates reduce rubber viscosity and cause it to flow faster for a given pressure change.

Temperature causes considerably less change in viscosity during molding than does the change in shear rate. This is true only if no cross-linking occurs. If cross-linking occurs while rubber is flowing, viscosity increases sharply and flow will virtually stop. Thus, a mold cavity should be filled completely with rubber before cross-linking initiates.

Compression molding is typically done over a shear rate range of about 1 to 10/s. Injection molding is usually done in a range of about 1000 to 10,000/s. Transfer molding is accomplished in a range intermediate between compression and injection. The viscosity of a rubber may be more than 100 times lower when it is being injection molded compared to when it is being compression molded. Thus, the rate at which the rubber moves, i.e., the shear rate, critically affects the ease with which rubber flows during molding. The rate of flow during compression molding is similar to the flow or shear rate that occurs in a Mooney viscometer, about 1/s. Hence, the Mooney

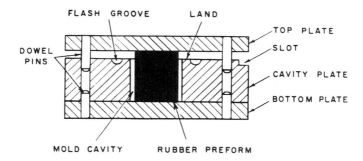

FIGURE 32.   A compression mold continuing rubber preform before closing the mold.

viscometer should provide viscosity data relevant to compression molding. Because injection molding is done at considerably higher rates, the Mooney viscometer is a poorer predictor of injection molding behavior. For this reason, the viscosity of rubber compounds for injection-molded rubber is sometimes measured at injection shear rates by high-rate viscometers. These are better able to predict viscosity and therefore injection-molding behavior.

Compression molds vary in size, shape and complexity, and can contain well over 300 cavities. A simple, single-cavity compression mold is shown in Figure 32. A cylindrical, uncured rubber preform is shown in its cavity prior to mold closure. The rubber preform is shaped to approximately fit the cavity. Because the preform does not fit the cavity exactly, it will keep the top plate above the cavity plate. During mold closure, rubber is squeezed between the top plate and the land and the excess flows into the flash groove. Typically, the volume of the flash groove is 10 to 20% of cavity volume. If the rubber preform is too large, excess rubber will flow past the flash groove toward the dowel pins. If the rubber gets into the clearance between the dowel pins and the hole into which they fit, mold opening and closing becomes difficult. Thus, the dowel pins should be sufficiently remote from the flash groove to avoid contact of rubber with the dowel pins. The purpose of dowel pins is to obtain proper register among mold plates and to prevent the mold plates from being assembled improperly. To prevent improper assembly, the dowel pins can be arranged so the mold can be assembled and closed only in the desired manner.

A molding press provides the force required to close the mold. A closed mold is shown in Figure 33. Excess rubber flows past the land into the flash groove. The thickness of flash above the land and the volume of flash in the flash groove will vary. The variable thickness flash between the land and top plate causes closure dimensions that are less accurate than fixed dimensions for compression molded parts. After the molding cycle is completed, the mold is opened. Opening is aided by placing a pry bar in the slot in the cavity plate.

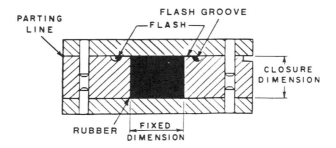

FIGURE 33.   A compression mold after closing.

The need for a separate cavity plate depends on the shape of the rubber part being molded. As the height/width ratio of a molded part increases, part removal from the mold becomes more difficult. For example, the rubber cylinder in Figure 33 can be pushed from the cavity plate only after both top and bottom plates are removed from the cavity plate. A three-plate mold is required because of the large area of contact between the walls of the rubber cylinder and the mold cavity wall.

A thin rubber sheet can be readily molded in a two-plate mold. There is relatively little surface area of the molded part in contact with the cavity walls. Hence, the cavity plate and bottom plate could be a single metal plate rather than two separate plates.

Small scale compression molds can be manually slid in and out of a press (as illustrated in Figure 34). As size increases, a mold eventually becomes too large and heavy for a press operator to manipulate. Large molds can be attached to press platens so that they are opened and closed by the press. Normally molds are heated by direct contact with press platens. Some molds are heated by steam or hot fluid and circulates through channels in a mold.

Considerable force is necessary to close a mold containing high viscosity rubber. For the majority of nontire or mechanical molded parts, hydraulic pressure is used. The two major types of molding press are the four-post press and the sideplate press. A schematic diagram of a four-post press with a mold in place is shown in Figure 34.

The role of the platens is to heat the mold uniformly. The source of this heat is generally steam, but hot fluid or electric heating are also used. With electric, resistance heating is most common, but induction heating is also used. To minimize undesirable heating of the ram bolster, ram, and press head, insulation is used between the lower platen and ram bolster and between the upper platen and the press head. This insulation is especially important for the ram bolster. If the ram bolster becomes too hot, heat transfers through the ram into the gland and hydraulic fluid. This heat accelerates deterioration of the packing gland and the hydraulic fluid in the cylinder.

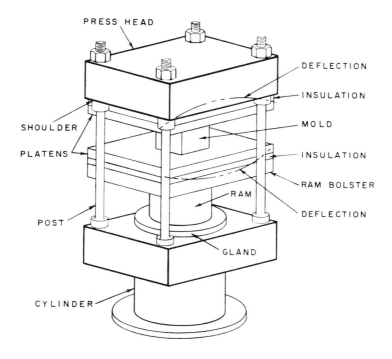

FIGURE 34.   A four-post molding press.

Another press design is the "C" frame which allows unobstructed access to the daylight (the gap or distance between the top of the ram bolster and the bottom of the press head) from the front and both sides.

Transfer molding is a variation of compression molding. In transfer molding, a plunger compresses a rubber preform in a pot, as shown in Figure 35. The rubber is heated by contact with the plunger face and pot. When sufficient force is applied to a mold by a press, rubber flows through the spur and into the mold cavity, as illustrated in Figure 36.

An important feature of a transfer mold is the depth of engagement of the plunger and pot. The plunger should engage the pot deeply enough to minimize tilting of the plunger in the pot. Another important consideration is the ratio between the area of the plunger face and the projected area of the cavity or cavities. The area of the plunger face must be larger than this projected area. If it is not, flash will form between the cavity plate and the plate containing the transfer pot. To avoid this the area of the plunger face is made about 30 to 50% larger than the projected cavity area.

Major differences exist among compression, transfer, and injection molding. With compression and transfer, it is necessary to place a rubber preform into a compression mold cavity or into a transfer pot. A strip of rubber, or granulated rubber, automatically supplies an injection-molding

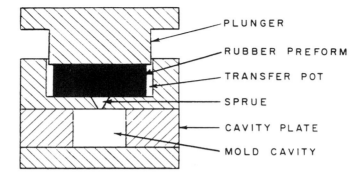

FIGURE 35. Transfer molding showing preform in transfer pot prior to mold closing.

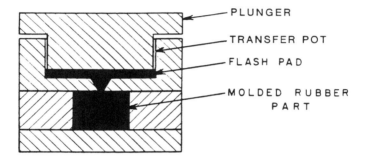

FIGURE 36. A transfer mold after closing.

machine. Another difference is that systems for injection molding are much more complex than those for compression or transfer molding; there are several controls to adjust temperature, pressure and other variables during injection molding. These controls are not normally part of compression and transfer molding systems.

With compression and transfer molding, presses provide the force to close a mold. In injection molding, a press is referred to as a clamp. A clamp is an integral part of an injection-molding machine. Injection molds are normally attached to the clamp and thus open and close with the clamp. One reason for attachment between mold and clamp is the need for accurate alignment between an injection molding machine and its mold. Injection molds must be capable of withstanding extremely high pressures without mold distortion. The pressure reached in an injection mold is about ten times greater than that for compression and transfer, i.e., injection pressures up to about 200 MPa (29,000 psi) are used.

In operation, preheated rubber from a nozzle on an injection-molding machine enters the sprue of a hot injection mold. For multicavity molds, the

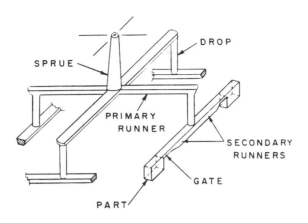

FIGURE 37.   A distribution system for an injection mold.

distribution system for this rubber is complex. A system for a multicavity mold is shown in Figure 37. Rubber flows through the sprue and enters four primary runners. These runners are trapezoidal in cross-section, but they might also be half-round or some other shape. The primary runner flows into a drop, which terminates at the secondary runner. The rubber divides in the secondary runner to feed the gates. A gate is the final restrictive pathway into a mold cavity.

A ram-type injection machine is shown in Figure 38 attached to a two-plate, single-cavity mold. The clamp used to keep the mold closed during molding is not shown. Heating fluid circulates in the barrel of the ram machine. Rubber is fed into its throat with the ram in the retracted (left) position. During injection, this ram forces rubber forward (to the right) into the injection chamber, through the nozzle and mold sprue where it then enters the mold cavity. Before rubber enters the mold cavity, it has been heated by contact with the inner barrel wall. Then the rubber becomes hotter as it passes through the nozzle and sprue at very high shear rates.

For the simple mold the sprue also serves as a gate. When a separate gate is used, rubber temperature rises significantly as rubber passes through the gate. The gate is usually small and restricts the rubber flow.

A reciprocating screw machine is shown in Figure 39. Its throat, barrel, injection chamber, nozzle, heating fluid, and mold are nearly identical to counterparts in the ram machine. The main difference is that the reciprocating screw machine is fitted with a screw which rotates inside the barrel. This screw not only rotates, it also moves back and forth on its axis like the ram design.

Rubber is fed into the throat of a reciprocating screw machine, where it then contacts the rotating screw. The screw plasticizes (softens) the rubber and transports or pumps it to the front of the injection chamber. Rubber

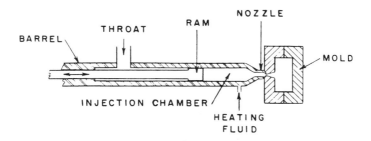

FIGURE 38. A ram-type machine for injection molding.

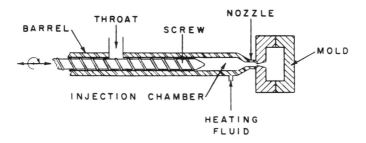

FIGURE 39. A reciprocating screw machine for injection molding.

accumulates there because the screw retracts (moves to left) to provide the needed volume in the injection chamber. When a sufficient volume of rubber accumulates, the screw moves toward the mold and fills the mold with rubber. Thus, the screw acts as both a pump and a ram. The screw develops less pressure and does not meter rubber as accurately as a ram. These undesirable features occur because the clearance between ram and barrel wall is typically less than between screw and barrel wall. With this small clearance, less leakage of rubber occurs during injection; hence, higher injection pressure and more accurate shots are obtainable with the ram-type machine, compared to the screw-type machine.

The reciprocating screw does work on the rubber in the barrel and raises rubber temperature above the barrel temperature; the higher temperature reduces rubber viscosity. This capability of the reciprocating screw machine has been combined with the higher injection pressure capability of the ram machine. Figure 40 shows a machine with a separate screw and ram. In this design the screw does not move back and forth on its axis; it only rotates, heating and pumping rubber to a three-way valve. The heated rubber flows through the valve into the injection chamber. A ram then forces rubber from the injection chamber into a mold. Hence, the best features of a ram machine and a reciprocating screw machine are combined in a single machine.

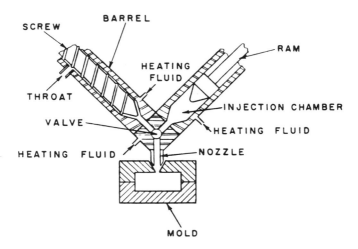

FIGURE 40. An injection-molding machine equipped with a separate screw and ram.

Arrangements of the components of an injection-molding machine fall into three categories: horizontal, vertical, and multiple-station rotary. The reciprocating screw-type machine is an example of a horizontal machine. A vertical machine is shown in Figure 40 (screw/ram machine). A vertical machine can be equipped with a sliding plate which permits moving the lower half of the mold to the operator; this movement facilitates removal of rubber parts and loading of inserts into the mold by the operator.

Another arrangement is the multiple-station rotary, an example of which is shown in Figure 41. This automated machine is used for molding shaft seals that contain an insert. In operation, rubber injects into a mold at station one and the table then rotates counterclockwise. Parts are demolded at station two, followed by mold cleaning at station three. Then metal inserts are loaded into the molds at station four and the cycle is repeated.

Injection molding can be combined with other molding techniques, e.g., compression or transfer. With injection/compression, an injection machine pumps hot rubber into an open compression mold. The mold is then closed, generally using much less force to close it than is normally used in injection molding.

Bladder molding is used for several different product systems. Basically it is a special case of compression molding. In fire molding a bladder inside an uncured tire pushes the exterior of the tire against a hot mold surface. This action forms the tread and sidewall patterns on a tire. Typically, a tire mold is made from steel or aluminum. Most molds have a circumferential parting line which is on-center. Some molds have a parting line that is slightly off-center to accommodate complex tread patterns. Figure 42 shows the cross-section of a tire in a curing press with a bladder in place. This figure also

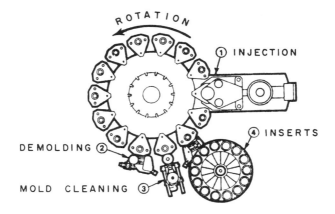

FIGURE 41. Multistation machine for injection-molding shaft seals containing an insert.

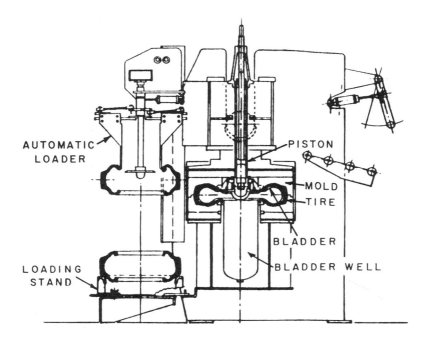

FIGURE 42. A tire curing press with tire in mold.

shows the loading stand used to store an uncured tire, along with the automatic loader which transfers tires from the loading stand to the mold. The automatic loader places an uncured tire in the open mold. Then the mold closes and a piston raises the bladder from the bladder well; low pressure steam forces the bladder into the uncured tire. Steam or hot water is then introduced into the

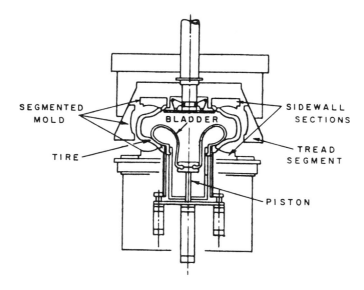

FIGURE 43.   A curing press containing a segmented mold.

bladder under high pressure. This pressure forces the tire into a mold. Thus, the tire is heated from the inside by the bladder and from the outside by the mold. After the tire is cured, the piston pushes the bladder downward and returns it to the bladder well. The cured tire is then removed automatically from the open mold.

Figure 43 illustrates a segmented mold in place in a tire press. The two sidewall sections of the segmented mold close to the final cure position before tread segments move toward the tire. These segments move into the tread area in exact symmetry with one another. Automatically controlled mechanical linkages accomplish the desired movement. In transitioning from the bladder well into the tire, the bladder turns inside out. With another design of tire curing press, the bladder does not do this. Instead, a center post inserts and removes the bladder from the tire. A vacuum on the bladder reduces its diameter so that the bladder can be inserted in the opening available in a tire. Once there, pressure is applied to the bladder during molding, as described above.

The equipment and methods employed to mold low viscosity (liquid-like) materials differ substantially from those used for high viscosity materials. Low pressure is used to mold liquid materials. Molds for liquid materials are lighter and sometimes made from nonmetallic materials such as silicone rubber or epoxy resin. Among liquid materials for molding, polyurethane, or simply urethane, is most common. The simplest urethane system consists of an isocyanate and a polyol blend. These components start reacting after they are mixed, causing the viscosity of the mix to increase. The rate of increase of viscosity for hand mixes is usually slow enough for them to be cast into a mold before increasing viscosity becomes a problem. Generally this time is several minutes, however, it depends upon the amount of material to be cast, reaction rate, complexity of the mold, number of cavities, and other factors.

Casting is the simplest method used to mold liquid materials. Reactants can be mixed by hand and then cast into a mold and cross-linked. Care must be taken during hand mixing and casting to minimize air entrapment. Air causes voids or bubbles in cast parts. For casting large parts or many small parts, a machine is usually used to mix reactants. Machines mix large quantities quickly and provide more uniform mixing than hand mixings. Roller skate wheels are an example of a molded urethane part made at high production rates by machine mixing. The mix from the machine is cast directly into open molds through a flexible tube using shot sizes accurate to 1 g. Large parts can be made by both hand and machine casting. If a sufficient amount of material cannot be mixed at one time to fill a large mold, multiple casts can be made. This procedure requires making rapid and successive casts so that the preceding cast does not cross-link to too great an extent, otherwise poor bonding will probably occur between the materials that were cast separately.

If a cast part contains inserts, trapping of air is a likely problem. This problem is more severe if a textile insert is used in a molded part; textiles contain many small pores or voids. When inserts such as these are present, vacuum centrifugal casting can be used to advantage. With this technique, the vacuum pulled on the casting drum removes air from the textile. This, of course, makes it much easier for the cast liquid to penetrate the textile and minimizes the occurrence of voids in the final composite.

Reaction injection molding (RIM) is used for mixing and molding liquid materials. A major difference between casting and RIM is the extremely rapid reaction rate for the reactants used to make RIM parts. This high reaction rate requires very rapid transfer of liquid from the mixing chamber to the mold. The process is shown in Figure 44. The polyol blend and isocyanate streams are under high pressure, about 13.8 MPa (2000 psi). The two streams enter the mixing chamber where they strike one another at very high velocities, causing turbulence. This turbulence results in thorough mixing of the streams, and the mixture then flows into the runner at low pressure. From the runner, the mixture flows through the gate into the mold cavity to form the molded part.

Because of the complexity of many parts, the high-precision requirements and the mass production needs of injection molders, robotics automation equipment has become popular in the plastics industry. Figure 45 shows one such unit manufactured by the Conair Group. This design features three axes with an AC servo motor drive having two additional slots in the controller for the possibility of five servo axes. For example, a typical application could include removal of part, palletizing, degating the sprue, and depositing into a grinder, and can also be programmed to periodically place a part on the

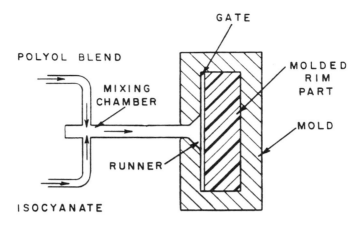

FIGURE 44.   Schematic of RIM.

opposite side of the press for quality control purposes — all in one program setup. All axes can be run simultaneously for greater speed when necessary, and with servo control, accuracy can be defined to 0.004 in.

The control panel is expandable and has large capacity for programmed motions within the mechanical reach of the robot. The controller has built-in subroutines for palletizing and can store up to 99 individual programs that can be called up as required within 30 s. This makes changeover extremely easy with the end-of-arm tooling change being the only mechanical work required.

In the rubber industry computer-integrated manufacturing is well established. This is exemplified in rubber injection-type applications. In molding operations one must remember that it is important to continuously check all production parameters to conform to the quality specifications of SPC (statistical process control) set by customers. Operations must be supervised and the flow of production coordinated to guarantee machine information flow (i.e., computer-integrated manufacturing) in order to obtain maximum efficiency and regulate lead times, while insuring quality parts of an improved profit level. Statistical process control is designed to guarantee consistent quality by defining for each piece part run the precise conditions needed to meet specifications and by ensuring that such conditions are met on every cycle of every press. A process can be considered to be under SPC when it allows:

- Real time control over factors of production by detecting variations from permitted tolerances
- The correction of recorded variations, thereby maintaining the process within specification

FIGURE 45. Total servo robot used for injection molding applications. (Courtesy of the Conair Group, Franklin Park Corporate Centre, 2000 Corporate Drive, Wexford, PA 15090. With permission.)

Commercial closed loop microprocessor control responds to these criteria by providing presses with the following capabilities by following critical quality control information to a central computer:

- Production times (curing, demolding, extrusion, cycle duration)
- Temperature of the injection unit, cold runner, and mold
- Extruder rotation speed and injection speeds
- Holding injection pressures, mold pressure, clamping pressure, back pressure, and average pressure during dynamic phase of injection
- Shot size

FIGURE 46.   Rubber injection molding machines. (Courtesy of REP Corp., P.O. Box 8146, 8N470 Tameling Court, Bartlett, IL 60103-8146. With permission.)

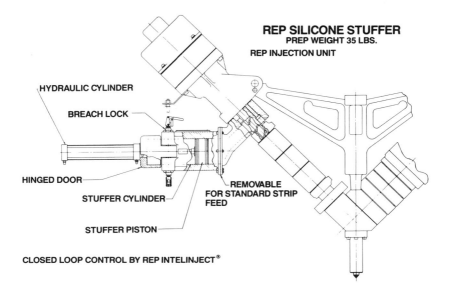

FIGURE 47. Rubber injection device. (Courtesy of REP Corp., P.O. Box 8146, 8N470 Tameling Court, Bartlett, IL 60103-8146. With permission.)

Photographs of typical rubber injection molding machines are shown in Figure 46. For illustration, a parts drawing of a "silicone stuffer" is shown in Figure 47. Finally, Figure 48 illustrates a typical clamping arrangement for a rubber mold. With the use of proper temperature sensing/control probes and sophisticated software, parts defection are dramatically reduced.

VI. ULTRASONIC PARTS ASSEMBLY

This last section provides an overview of polymer welding techniques based on ultrasonics. Polymer welding techniques have proven highly successful in the mating and finishing of thermoplastics parts. Although it is somewhat esoteric with reference to applied processing rheology, its importance in overall process operations is noteworthy.

Ultrasonic assembly is a fast, clean, efficient method of assembling or processing rigid thermoplastic parts or films as well as synthetic fabrics. A variety of ultrasonic assembly techniques are employed by different segments of industry to join plastic and plastic to metal parts or other nonplastic materials, replacing or precluding the use of solvents, adhesives, mechanical fasteners, or other consumables.

A typical assembly converts 50/60 Hz current to 20 or 40 kHz electrical energy through a solid-state power supply. This high frequency electrical energy is supplied to a converter, a component that changes electrical energy into mechanical vibratory energy at ultrasonic frequencies. The vibratory

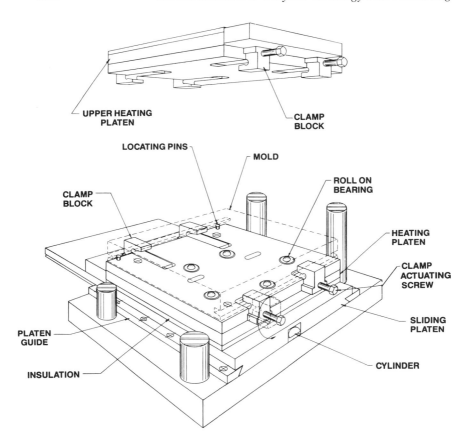

FIGURE 48.   Rubber mold. (Courtesy of REP Corp., P.O. Box 8146, 8N470 Tameling Court, Bartlett, IL 60103-8146. With permission.)

energy is then transmitted through an amplitude-modifying device (referred to as a booster) to the horn. The horn is an acoustic tool that transfers this vibratory energy directly to the parts being assembled.

The vibrations are transmitted through the workpiece to the joint area where vibratory energy is converted to heat through friction that melts the plastic. When this molten state is reached at the past interface, vibration is stopped; pressure is maintained briefly on the parts while the molten plastic solidifies, creating a strong molecular bond between the parts. Cycle times are usually <1 s, and weld strength obtained approaches that of the parent material.

Ultrasonic vibratory energy is used in several distinct assembly and finishing techniques as described below.

Welding — The process of generating melt at the mating surfaces of two thermoplastic parts. When ultrasonic vibrations stop, the molten material solidifies and a weld is achieved. The resultant joint strength approaches that of the parent material; with proper part and joint design, hermetic seals are possible. Ultrasonic welding allows fast, clean assembly without the use of consumables.

Staking — The process of melting and reforming a thermoplastic stud to mechanically lock a dissimilar material in place. Short cycle times, tight assemblies, good appearance of final assembly, and elimination of consumables are possible with this method.

Inserting — Embedding a metal component (such as a threaded insert) in a preformed hole in a thermoplastic part. High strength, reduced molding cycles, and rapid installation with no stress buildup are some of the advantages.

Swaging/forming — Mechanically capturing another component of an assembly by ultrasonically melting and reforming a ridge of plastic or reforming plastic tubing or other extruded parts. Advantages of this method include speed of processing, less stress buildup, good appearance, and the ability to overcome material memory.

Spot welding — Ultrasonic spot welding is an assembly technique for joining two thermoplastic components at localized points without the necessity for preformed holes or an energy director. Spot welding produces a strong structural weld and is particularly suitable for large parts, sheets of extruded or cast thermoplastic, and parts with complicated geometry and hard-to-reach joining surfaces.

Slitting — The use of ultrasonic energy to slit and edge-seat knitted, woven, and nonwoven thermoplastic materials. Smooth, sealed edges that will not unravel are possible with this method. There is no "bead" or buildup of thickness on the slit edge to add bulk to rolled materials.

Textile/film sealing — Fabric and film sealing utilizes ultrasonic energy to join thin thermoplastic materials. Clear, pressure-tight seals in films, and neat, localized welds in textiles may be accomplished. Simultaneous cutting and sealing is also possible.

It must be remembered that a polymer is repeating structural unit formed during a process called polymerization. There are two basic polymer families: thermoplastic and thermoset. A thermoplastic material, after being formed can, with the reintroduction of heat and pressure, be remelted and reformed, undergoing only a change of state. This characteristic makes thermoplastics suitable for ultrasonic assembly. A thermoset is a material that once formed undergoes an irreversible chemical change and cannot be reformed with the reintroduction of heat and pressure; therefore, thermosets cannot be ultrasonically assembled in the traditional sense.

When discussing the weldability of thermoplastics, it must be recognized that there are a number of factors that affect the ultrasonic energy requirements and, therefore, weldability of the various resins. The major factors include polymer structure, melt temperature, modulus of elasticity (stiffness), and chemical makeup.

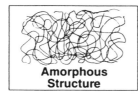 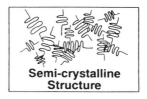

FIGURE 49.   Structural arrangement of amorphous and semi-crystalline polymers.

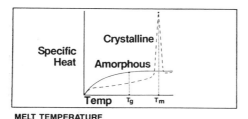

FIGURE 50.   Specific heat vs. temperature.

Amorphous resins are characterized by a random molecular arrangement and a broad softening temperature (T_g, glass transition temperature) range that allows the material to soften gradually, melt, and flow without prematurely solidifying. These resins generally are very efficient with regard to their ability to transmit ultrasonic vibrations, and can be welded under a wide range of force/amplitude combinations.

Semi-crystalline resins are characterized by regions of orderly molecular arrangement and sharp melting (T_m, melt temperature) and resolidification points (refer to Figure 49). The molecules of the resin in the solid state are spring-like and internally absorb a percentage of the high frequency mechanical vibrations, thus making it more difficult to transmit the ultrasonic energy to the joint interface. For this reason, high amplitude is usually required. The sharp melting point is the result of a very high energy requirement (high heat of fusion) necessary to break down the semicrystalline structure to allow material flow. Once the molten material leaves the heated area, these resins solidify rapidly with only a small reduction in temperature. These characteristics therefore warrant special consideration (i.e., higher amplitude, careful attention to joint design, horn contact, and fixturing) to obtain successful results.

The higher the melt temperature of a resin, the more ultrasonic energy is required for welding (Figure 50).

The stiffness of the resin to be welded can influence its ability to transmit the ultrasonic energy to the joint interface. Generally the stiffer a material the better its transmission capability.

A similar melt temperature between the materials to be welded is a basic requirement for successful welding of rigid parts, because a temperature difference of 40°F (22°C) can be sufficient to hinder weldability (even for a like resin). The lower melt temperature material melts and flows, preventing the generation of sufficient heat to melt the higher melt temperature material. For example, with an energy director on a part composed of high temperature acrylic, the weld surface of the high temperature part will not reach the necessary temperature to melt. The opposing surface will be in a molten state before the energy director begins to soften, and if the energy director fails to melt, bond strength will be impossible to predict.

In addition, to weld dissimilar plastics, the plastics to be welded must possess a like molecular structure (i.e., be chemically compatible) with some component of the material, usually a blend. Close examination of compatible thermoplastics reveals that like radicals are present, and the percentage of the like chemical radical will determine the molecular compatibility and bond strength. *Note:* Compatibility exists only among amorphous resins or blends containing amorphous resins. Refer to Table 1 for characteristics. The codes in this table indicate relative ease of welding for the more common thermoplastics. In addition to the material factors covered in the preceding discussions, ease of welding is a function of joint design, part geometry, energy requirements, amplitude, and fixturing. *Note:* The ratings in the table do not relate to the strength of the weld obtainable. Some materials are hygroscopic, i.e., they absorb moisture which can seriously affect weld quality. Nylon (and to a much lesser degree polyester, polycarbonate, and polysulfone) is the material most troubled by these characteristics. If hygroscopic parts are allowed to absorb moisture, when welded the water will evaporate at 212°F (100°C), with the trapped gas creating porosity (foamy condition) and often degrading the resin at the joint interface. This results in difficulty in obtaining a hermetic seal, poor cosmetic appearance (frostiness), degradation, and reduced weld strength. For these reasons, if possible it is suggested that nylon parts be welded directly from the molding machine to insure repeatable results. If welding cannot be done immediately, parts should be kept dry-as-molded by sealing them in polyethylene bags or other suitable means directly after molding. Drying of the parts prior to welding can be done in special ovens; however, care must be taken to avoid material degradation.

Using additives or processing aids during the preparation of a resin compound may result in properties not inherent in the base resin. These additives, which can enhance certain areas of processing, can in some cases create problems in ultrasonic welding. Molding release agents, often called parting agents, are applied to the surface of the mold cavity to provide a release coating which facilitates removal of the parts. External release agents, such as zinc stearate, aluminum stearate, fluorocarbons, and silicones can be transferred to the joint interface and interfere with surface heat generation and fusion, inhibiting welding; silicones are generally the most detrimental. If it

TABLE 1
Characteristics of Amorphous and Semi-crystalline Resins

Material	Ease of Welding		Swaging and Staking	Insertion	Spot Welding	Vibration Welding
	Near Field*	Far Field*				
Amorphous Resins						
ABS	E	G	E	E	E	E
ABS/polycarbonate alloy	E-G	G	G	E-G	G	E
Acrylic[a]	G	G-F	F	G	G	E
Acrylic multipolymer	G	F	G	G	G	E
Butadiene-styrene	G	F	G	G	G	G
Phenylene-oxide based resins	G	G	G-E	E	G	E-F
Polyamide-imide	G	F				G
Polyarylate	G	F				
Polycarbonate[b]	G	G	G-F	G	G	E
Polyetherimide	G	F				
Polyethersulfone	G	F				
Polystyrene (general purpose)	E	E	F	G-E	F	E
Polystyrene (rubber modified)	G	G-F	E	E	E	E
Polysulfone[b]	G	F	G-F	G	F	E
PVC (rigid)	G-F	P	G	E	G-F	G
SAN-NAS-ASA	E	E	F	G	G-F	E
Xenoy (PBT/polycarbonate alloy)	G	F	F	G	G	E

Material	Ease of Welding		Swaging and Staking	Insertion	Spot Welding	Vibration Welding
	Near Field*	Far Field*				
Semi-Crystalline Resins[c]						
Acetal	G	F	G-F	G	F	E
Cellulosics	F-P	P	G	E	F-P	E
Fluoropolymers	P					F
Ionomer	F	P				
Liquid crystal polymers	F	P	G-F			
Nylon[b]	G	F	G-F	G	F	E
Polyester, thermoplastic						
Polyethylene terephthalate-PET	G-F	P				
Polybutylene terephthalate-PBT		P				
Polyetheretherketone-PEEK	F	P				G
Polyethylene	F-P	P	G-F	G	G	G-F
Polymethylpentene	F	F-P	G-F	E	G	E
Polyphenylene sulfide	G	F	P	G	F	G
Polypropylene	F	P	E	G	E	E

Code: E = Excellent, G = Good, F = Fair, P = Poor

* Near field welding refers to a joint 1/4 in. (6.35 mm) or less from the horn contact surface; far field welding refers to a joint more than 1/4 in. (6.35 mm) from the horn contact surface.

[a] Cast grades are more difficult to weld due to higher molecular weight.
[b] Moisture will inhibit welds.
[c] Semi-crystalline resins in general require higher amplitudes due to polymer structure and higher energy levels due to higher melt temperatures and heat of fusion.

Courtesy of Branson Ultrasonic Corp., 41 Eagle Road, Danbury, CT 06813-1961. With permission.

is absolutely necessary to use an external release agent, the paintable/printable (nontransferring) grades should be used. These grades prevent the resin from wetting the surface of the mold, with no transfer to the molded part itself, thus permitting painting and silk-screening and the least amount of interference with ultrasonic assembly. Detrimental release agents can in some cases be removed by using a solvent, such as TF Freon. Internal molded-in release agents, since they are generally uniformly dispersed internally in the resin, usually have minimal effect on the welding process.

Lubricants (internal and external) are materials that enhance the movement of the polymer against itself or against other materials (examples include waxes, zinc, stearate, stearic acid, esters). Lubricants reduce intermolecular friction (melt viscosity) within the polymer or reduce melt flow friction against primary processing equipment surfaces. Since molecular friction is a basis for ultrasonically induced temperature elevation, lubricants can inhibit the ultrasonic assembly process. However, since they are generally dispersed internally, like internal mold release agents their effect is usually minimal.

Plasticizers are high-temperature boiling organic liquids or low-temperature melting solids that are added to resins to impart flexibility. They do this through their ability to reduce the intermolecular attractive forces of the polymer matrix. They can also interfere with the ability of a resin to transmit vibratory energy. Attempting to transmit ultrasonic vibrations through a highly plasticized material (such as vinyl) is like transmitting energy through a sponge. Even though plasticizers are considered an internal additive, they do migrate to the surface over time, and the combination of internal as well as surface lubricity make plasticized vinyl all but impossible to weld. Food and Drug Administration-approved plasticizers do not present as much of a problem as metallic plasticizers, but experimentation is recommended. Table 2 provides comparative data on resin compatibility with various thermoplastics.

Impact modifiers such as rubber can affect the weldability of a material by reducing the amount of thermoplastic available at the joint interface. They can also reduce the ability of the resin to transmit ultrasonic vibrations, making it necessary to increase amplitude to generate a melt.

Foaming agents also reduce the ability of a resin to transmit energy. Voids in the cellular structure interrupt the energy flow, reducing the amount of energy reaching the joint area, depending on the density.

Flame retardants are added to a resin to inhibit ignition or modify the burning characteristics. They can adversely affect ultrasonic welding characteristics of the resin compound. Flame-retardant chemicals are generally inorganic oxides or halogenated organic elements, and for the most part are nonweldable. Typical examples are aluminum, antimony, boron, chlorine, bromine, sulfur, nitrogen, and phosphorus. The amount of flame-retardant material required to meet certain test requirements may vary from a few percent to 50% or more by weight of the total matrix, thus reducing the amount of available weldable material. This reduction must be compensated

TABLE 2
Compatibility of Thermoplastics

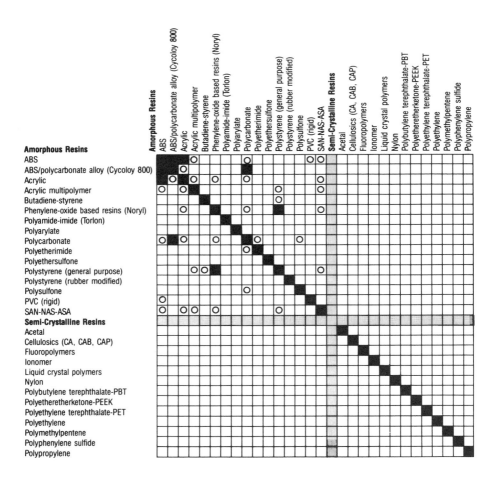

■ – Denotes compatibility
O – Denotes compatibility in some cases (usually blends)

Courtesy of Branson Ultrasonic Corporation, 41 Eagle Road, Danbury, CT 06813-1961. With permission.

for by modifying the joint configuration to increase the amount of weldable material at the joint interface and by increasing ultrasonic energy levels.

Scrap formed during the molding process, e.g., sprues, runners, reject parts, can usually be recycled directly back into the process after the material has been reduced to a usable size. Control over the volume and quality of regrind is necessary, as it can adversely affect the welding characteristics of the molded part. In some cases the use of 100% virgin material may be required to obtain the desired results.

Most colorants, either pigments or dyestuffs, do not interfere with ultrasonic assembly; however, occasionally some pigments (white, black) can influence weldability. Titanium dioxide (TiO_2) is the main pigment used in white parts. TiO_2 is inorganic, chemically inert, and can act as a lubricant, and if used in high loadings (>5%), it can inhibit weldability. Black parts, on the other hand, can be pigmented with carbon, which can also inhibit weldability. In any event, an application evaluation should be undertaken. Parts molded in different pigments may require minor variations in welding parameters.

Resin grade can have a significant influence on weldability because of melt temperature and other property differences. An example is the difference between injection/extrusion grades and cast grades of acrylic. The cast grade has a higher molecular weight and melt temperature and other property differences. An example is the difference between injection/extrusion grades and cast grades of acrylic. The cast grade has a higher molecular weight and melt temperature, is often brittle, and forms a skin that gives it greater surface hardness, all of which reduce weldability to the injection grade. A general rule of thumb is that both materials to be welded should have similar molecular weight, and melt temperatures within 40°F (22°C) of each other.

Filled or reinforced plastics, in which additives, fillers, and reinforcements are combined with the base resin, are widely known as composites. Advanced composites refer to high-performance fibers such as graphite or Kevlar, and high-performance materials formed by means of more complex techniques involving weaving, winding, or otherwise aligning reinforcement into special patterns. The following information deals only with composites.

Fillers/extenders constitute a category of additives (nonmetallic minerals, metallic powders, and other organic materials) added to a resin that alter the physical properties of resins. Fillers enhance the ability of some resins to transmit ultrasonic energy by imparting higher rigidity (stiffness). Common materials such as calcium carbonate, kaolin, talc, alumina trihydrate, organic filler, silica, glass spheres, wollastonite (calcium metasilicate), and micas can increase the weldability of the resin considerably; however, it is very important to recognize that a direct ratio between the percentage of fillers and the improvement of weldability exists only within a predescribed quantitative range. Up to 20% can actually enhance weldability, due to increased stiffness, giving better transmission of the vibratory energy to the joint.

Resins with a filler content up to 10% can be welded in a normal manner, without special procedures and equipment. However, with many fillers when filler content exceeds 10% the presence of abrasive particles at the resin surface can cause horn wear. In this situation the use of hardened steel or carbide-faced (coated) titanium horns is recommended.

When filler content approaches 35%, there may be insufficient resin at the joint surface to obtain reliable hermetic seals; when filler content exceeds 40%, tracking, or the accumulation of filler (typically fibers), can become so severe that insufficient base resin is present at the joint interface to form a consistent bond.

It should be noted that particular types of fillers can present special problems. When long fibers of glass are employed, they can collect and cluster at the gate area during molding, being forced through in lumps rather than uniformly dispersed. This agglomeration can lead to an energy director containing a much higher percentage of glass. If this were to occur, no appreciable weld strength could be achieved since the energy director would embed itself in the adjoining surface, not providing the required molten resin to cover the joint area. If this problem occurs, it can be eliminated by utilizing short fiberglass filler.

Fibrous reinforcements of resins can, like fillers or extenders, be used to enhance or alter physical properties of the base resin. Continuous or chopped fiber strands of aramid, carbon, glass, etc., can in some cases improve the weldability of a resin; however, rules governing the use of fillers should be observed.

Ultrasonic energy has been used to join thermoplastics for over 25 years. Ultrasonic welding of thermoplastic materials is by far the most common form of ultrasonic assembly, and is used extensively in all major industries including automotive, appliance, electronic, toy, packaging, textile, and medical. The primary factors influencing joint design are: What type of material(s) is to be used? What are the final requirements of the part?

All of the following basic questions must be answered prior to the design stage to gain a total understanding of what the weld joint must do:

• Is a structural bond desired and, if so, what load forces does it need to resist?
• Is a hermetic seal required?
• Does the assembly require a visually attractive appearance?
• Is flash/particulate objectionable?
• Any other requirements?

In order to obtain acceptable, repeatable welded joints, three general design guidelines must be followed:

1. The initial contact area between the mating surfaces should be small to concentrate and decrease the total energy (and thus the time) needed to start and complete melting. Minimizing the time the vibrating horn remains in contact with the part also reduces the potential for scuffing, and since less material is moved, less flash occurs.

2. A means for aligning the mating parts should be provided. Features in part design such as pins and sockets, steps or tongues, and grooves should be used for alignment rather than the vibrating horn and/or fixture, to ensure proper, repeatable alignment and to avoid marking.
3. Mating surfaces around the entire joint interface should be uniform and in intimate contact with each other. If possible, the joint area should be on a single plane. Such measures enable the uniform transfer of energy for consistent (and therefore controllable) melting and welding, and a reduction in the potential for flash.

There are two major types of design: the energy director and the shear joint. All other joint variations can be classified under these general categories or as hybrids combining aspects of both. The energy director is typically a raised triangular bead of material molded on one of the joint surfaces. The primary function of the energy director is to concentrate the energy to rapidly initiate the softening and melting of the joining surface. The diagrams in Figure 51 show time-temperature curves for a common butt joint and the more ideal joint incorporating an energy director. The energy director permits rapid welding while achieving maximum strength; material within the director flows throughout the joint area. The energy director is the most generally used design for amorphous materials, although there can be deviations from this rule. The size and location of the energy director on the joint interface is dependent upon material(s), mold construction, and the requirements of the application.

Formulas for typical dimensions for an energy director are shown in Figure 52. When using a relatively easy-to-weld resin [e.g., polystyrene which has high modulus (stiffness) and low melt temperature] a minimum height of 0.010 in. (0.25 mm) is suggested, whereas a semicrystalline or high temperature amorphous resin (e.g., polycarbonate) would require a minimum height of 0.020 in. (0.5 mm). In the case of semicrystalline resins (e.g., acetal, nylon) with an energy director, the maximum joint strength is generally obtained only from the width of the base of the energy director.

As a rule, it does not make a difference which half of the part contains the energy director. In special situations (as in combinations of different materials), the general practice is to place the energy director on the part with the material that has the highest melt temperature and stiffness. The energy director design requires a means of alignment such as pins and sockets, aligning ribs, tongue and groove designs, or fixturing. Knockout pins should not be placed in the weld area.

The basic energy director design can be incorporated into joint configurations other than the butt joint to gain additional benefits. Examples of joint design variations utilizing an energy director include the following alternatives.

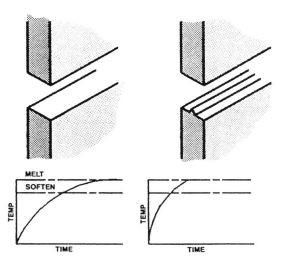

FIGURE 51.   Time-temperature curves for butt joints and ideal joint incorporating an energy director.

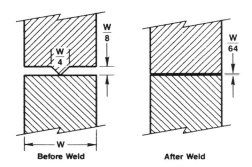

FIGURE 52.   Typical dimensions for an energy director.

Step joint — The step joint is used for alignment and for applications where excess melt or flash on one exposed surface is objectionable (Figure 53).

Tongue and groove — The major benefits of using this joint design are that it prevents flash both internally and externally and provides alignment. Containment of the material enhances the attainment of hermetic seals. The need to maintain clearance on both sides of the tongue, however, makes this more difficult to mold. Also, because of the reduce weld area, it is generally not as strong as a full butt joint (Figure 54).

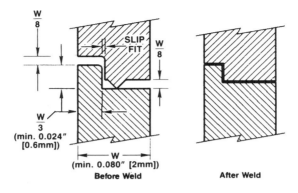

FIGURE 53. A step joint.

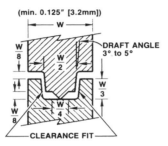

FIGURE 54. A tongue and groove joint.

Shear joint — An energy director type of joint design in some cases may not produce the desired results with semicrystalline resins such as nylon, acetal, polypropylene, polyethylene, and thermoplastic polyester. This is due to the fact that semicrystalline resins change rapidly from a solid to a molten state, or from the molten state to a solid, over a relatively narrow temperature range. The molten material flowing from an energy director, therefore, would rapidly resolidify before fusing with the adjoining interface. A shear joint configuration is recommended for these resins where geometry permits. With a shear joint design, welding is accomplished by first melting the small, initial contact area and then continuing to melt with a controlled interference along the vertical walls as the parts telescope together (refer to Figure 55). This allows a strong structural or hermetic seal to be obtained as the molten area of the interface is never allowed to come in contact with the surrounding air. For this reason, the shear joint is especially useful for semicrystalline resins.

The strength of the sealed joint is a function of the vertical dimension of meltdown of the joint (depth of weld), which can be adjusted to meet the requirements of the application. *Note:* For joint strength exceeding that of the wall, a depth of 1.25 × the wall thickness is recommended.

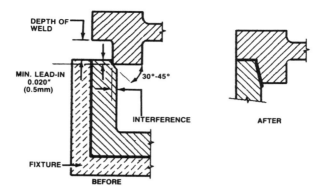

FIGURE 55.   A shear joint design.

Ultrasonic insertion — The process of embedding or encapsulating a metal component in a thermoplastic part. This process replaces the costly, time-consuming, conventional method of injection-molding plastic around the metal component (insert molding). An endless variety of part configurations can be inserted — flat, round, etc.; the most common is round, threaded inserts. This technical information sheet provides information on the fundamental principles of ultrasonic insertion, guidelines for insert and hole design, and basic rules for efficient use of the technique.

In ultrasonic insertion, a hole slightly smaller than the insert it is to receive is either molded or drilled into the plastic part. This hole provides a degree of interference (usually 0.015 to 0.020 in. in diameter) and guides the insert into place. The metal insert is usually designed with exterior knurls, undercuts, or threads to resist loads imposed on the finished assembly.

Ultrasonic insertion can be accomplished either by driving the insert into the plastic or by driving the plastic component over the metal insert (see Figure 56). Ultrasonic vibrations travel through the driven component to the interface of the metal and plastic. Frictional heat is generated by the metal insert vibrating against the plastic, causing a momentary, localized melting of the plastic. As the insert is driven into place, the molten material flows into the serrations and undercuts of the insert. When ultrasonic energy ceases, the plastic resolidifies, locking it in place.

In the majority of insertion applications the horn contacts the insert. However, several advantages exist to contacting the plastic part instead of the insert (inverse insert ion). The horn wear problem encountered when the horn contacts the metal insert is alleviated, allowing the use of aluminum chrome-plated or titanium horns. Also, the noise level is reduced, the power requirement is not as high, and a converter protection circuit in the power supply is not necessary.

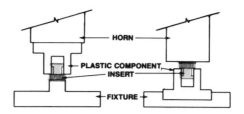

FIGURE 56. The process of ultrasonic insertion.

Ultrasonic insertion offers several advantages over the other insert assembly techniques including:

- Short cycle time — typically <1 s.
- No induced stress around the metal insert, as with molding-in or cold pressing.
- Elimination of possible mold damage and downtime should inserts fall into the mold during insert molding.
- Reduced molding cycle times.
- Multiple inserts can be driven at one time.
- Ideal for automated, high production operations.
- Repeatability and control over the process.

The insert and hole designs are determined by the functional characteristics or requirements of an application. A sufficient volume of plastic must be displaced to fill the exterior undercuts, knurls, and/or threads of the insert to lock it in place, and produce the strength required for the application.

A typical threaded insert designed primarily for tensile (pullout) strength should have multiple undercuts to provide maximum resistance (Figure 57). Where maximum torque strength is required, the insert should have long axial (straight) knurls. Combination inserts are designed with both undercuts and knurls to provide pullout and torque strength.

The insert acceptance hole should be designed so a sufficient volume of material is present to be displaced by the insert. This material will fill the exterior undercuts, knurls, and/or threads of the insert to lock it in place and provide the strength required for the application. Additionally, the acceptance hole should be designed so the insert does not bottom out; the recommended minimum depth of the hole is the length of the insert plus 0.030n, to provide a gap for forward-displaced material. This is especially important with threaded bore inserts, as it prevents material from being driven up through the bore, which could render the insert useless.

The hole should also be sufficiently deep to prevent a threaded screw from bottoming out in the final assembly. This also can be prevented by specification of the proper screw length. The hole may be tapered, especially if it is to accept a tapered insert; this facilitates accurate positioning of the insert and usually reduces installation time.

When inserting into a blind hole, a vent along the insert or acceptance hole should be provided. This prevents a pressure buildup under the insert, which could make it difficult to gain repeatable results and affect the appearance of the surface of the part around the insert.

A lead-in designed into the insert or the acceptance hole that allows 50% of the insert to be located in the hole prior to insertion is suggested. This provides easier handling of the parts and prevents the insert from moving out of the hole when initially contacted by the horn or during indexing in automation.

A small flange can be used on the top of an insert to create a larger horn contact area and also push material that may have been displaced up the sides of the insert back down around the sides.

Should a hermetic seal be desired, a specially designed insert that incorporates a gasket or ''O'' ring is required to achieve consistent results.

The majority of inserting applications involve the installation of standard threaded bore inserts; however, other metal components can be inserted, including eyeglass hinges, machine screws, threaded rods, metal bezels, roll pins, metal shafts, metal mesh or screens, decorative trim, electrical contacts, terminal connectors, fabrics, and higher melt-temperature plastics.

Ultrasonic stud welding — In many applications requiring permanent assembly, a continuous weld is not required. Frequently the size and complexity of the parts seriously limit the type of joint design that can be used. With compatible materials, this type of assembly can be effectively and economically accomplished using ultrasonic stud welding, a reliable assembly technique which can be used to join thermoplastic parts at a single point or numerous locations.

Ultrasonic stud welding is the process of joining thermoplastic parts at localized areas, singly or in multiples, by driving a molded stud into a hole with an interference fit. The vibrating horn contacts the outer surface of the part over the weld (stud) area, causing a frictional heat buildup to occur at the interface of the stud and socket, melting the thermoplastic material, and thus allowing the sections to telescope together. At the end of the weld cycle, the vibrations are stopped, and the parts are held together under pressure allowing the molten material to resolidify. As with any good ultrasonic welding joint, a part designed for stud welding must meet three basic requirements: small initial contact area, uniform contact, and means of alignment.

Figure 58 shows the basic stud welding joint before, during, and after welding. Welding occurs along the circumference of the stud and socket due to the interference. The radial interference must be uniform and should generally be 0.008 to 0.012 in. (0.2 to 0.3 mm) for studs having a diameter of 0.5 in. (12.5 mm) or less. This also provides the small initial contact area required, which helps reduce the cycle time by concentrating the vibratory energy in a localized area.

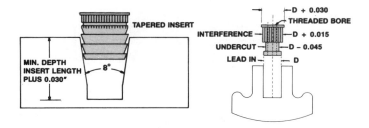

FIGURE 57. Insert and hold designs. (Left) Design for tensile (pullout) strength; (Right) long axial (straight) knurls.

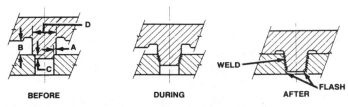

DIMENSION A: 0.008″ to 0.012″ (0.2 to 0.3 mm) for D up to 0.5 inches (12.5 mm).
DIMENSION B: Depth of weld. B=0.5 D for maximum strength (stud to break before joint failure).
DIMENSION C: 0.016 inches (0.4 mm) minimum lead-in. The step can be at the end of the pin or the top of the hole.
DIMENSION D: Stud diameter.

FIGURE 58. Ultrasonic stud welding joint.

The means of alignment is provided by the lead-in which can be on the end of the stud or at the top of the acceptance hole. When using the latter, a small chamber can be used for rapid alignment.

To reduce stress concentration during welding and in use, an ample fillet radius should be incorporated at the base of the stud. Recessing of the fillet below the surface serves as a flash trap which allows flush contact of the parts. The resulting strength of the weld is a function of the stud diameter and the depth of weld (i.e, total area). Maximum strength in tension is achieved when the depth of weld equals half the diameter. When appearance is important, or where an uninterrupted surface is required, welding a stud into a boss is an alternative design. The outside diameter of the boss should be no less than two times the stud diameter. The acceptance hole should be a sufficient distance from the edge to prevent deflection or breakout. A minimum of 0.125 in. (3 mm) is recommended (see Figure 59). Due to the possible pressure buildup which could result from welding into a blind hole, it may be necessary to provide an outlet for air. Two possibilities can be considered: a center hole through the stud or wall section, or a small narrow keyway in the boss.

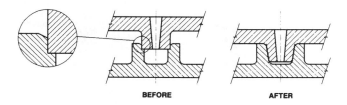

FIGURE 59. Stud in boss or bind hole.

Ultrasonic staking — In manufacturing products with thermoplastic components, it is often necessary to join a thermoplastic to a part of dissimilar material, whether it be metal, a dissimilar plastic, or other material. Ultrasonic staking is an assembly method that uses the controlled melting and reforming of a plastic stud or boss to capture or lock another component of an assembly in place. The plastic stud protrudes through a hole in the component in order to be locked in place. High frequency ultrasonic activity from the horn is imparted to the top of the stud, which melts and fills the volume of the horn cavity to produce a head, locking the component in place. The progressive melting of plastic under continuous but generally light pressure forms the head.

The advantages of ultrasonic staking include:

- Short cycle time (generally <1 s)
- Tight assemblies with virtually no tendency for recovery (memory)
- The ability to perform multiple stakes with one horn
- Repeatability and control over the process
- Design simplicity
- The elimination of consumables such as screws and rivets

In order to design the part correctly, a number of questions must first be answered:

1. What material is being used?
2. What strength will be required?
3. What loading must the stake resist in normal use (e.g., tensile, shear)?
4. Is appearance important?
5. Will multiple staking be necessary? If so, what is the distance between the studs?
6. Are the stakes recessed in the part and if so, is there clearance for the horn?

These questions must be answered to determine the requirements of the application, for it is these requirements and the physical size of the stud(s) being staked that determine the type of design to be utilized.

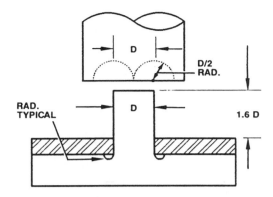

FIGURE 60.   The relationship between stud and horn cavity.

The integrity of an ultrasonically staked assembly depends upon the volumetric relationship between the stud and horn cavity and the ultrasonic parameters used when forming the stud (e.g., amplitude of the horn, weld time, pressure) (see Figure 60). Proper stake design produces optimum strength and appearance with minimal or no flash.

Several configurations for stud/cavity design are available. The principle of staking is the same for each: the area of initial contact between the horn and stud should be kept to a minimum. This allows a concentration of the mechanical vibrations in a localized area to create a rapid melt, which speeds up the cycle on the part. This is true with each of the following designs, as well as customized designs to meet a specific part requirement.

Standard profile stake — The standard profile stake (Figure 61) is most commonly used for studs having a diameter between 1/8 and 5/32 in. (3.2 to 4 mm). The top of the molded stud is flat, and melt is initiated by the small, extended point in the horn cavity. The head produced is twice the diameter of the stud and satisfies the requirements of the majority of staking applications. It is ideal for staking nonabrasive (unfilled) thermoplastics, both rigid and nonrigid. Standardized threaded horn tips for tapped horns are available for studs with diameters of 1/32 to 3/16 in. (0.8 to 4.8 mm). The standard profile should not be used for studs >5/32 in. (4 mm) in diameter. Low profile or hollow staking should be used.

Ultrasonic degating — The separation of injection-molded parts from their runner systems through the introduction of ultrasonic energy into the area of the gate is referred to as ultrasonic degating. These gate sections, when ultrasonically excited, reach high temperatures and melt due to a high degree of intermolecular excitation and internal friction. The advantages of ultrasonic degating include speed of operation, with a typical cycle time of <1 s; no stress developed in plastic component, and a clean break at the part surface. Many parts may be degated and automatically separated into bins.

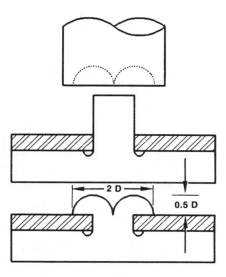

FIGURE 61. The standard profile stake.

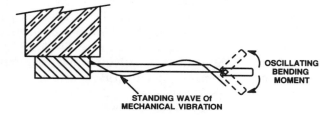

FIGURE 62. Ultrasonic degating.

The vibrating horn contacts the runner as closely to the part as possible with relatively light force, generally 20 to 40 lb, imparting maximum energy into the runner system. The mechanical vibrations generate a standing wave of energy down the runner through the gate area and into the part. (Refer to Figure 62.) The freely resonating part goes through cyclic bending moment with the safe area acting as the fulcrum. This bending induces stress into the already highly stressed gate area and generates internal molecular friction (hysteria losses) which rapidly elevates the temperature of the material at the localized area of the gate. Once the melt temperature is reached, the part falls free of the runner. What type of material is to be used?

• Rigid thermoplastic materials (polystyrene, ABS, polycarbonate, acrylic, nylon, etc.) work best. They transmit the mechanical energy more efficiently.

- Lower stiffness (modulus) thermoplastics such as polyethylene and polypropylene absorb the mechanical energy and do not give good consistent results.

What is the part size/mass?

- Generally parts of smaller size and mass are better suited for ultrasonic degating. Large, massive parts require excessive energy to overcome inertia.
- Brittle thermoplastics tend to fracture from the mechanical vibrations instead of melting. Thermosets, of course, do not melt at all, but a mechanical break is still possible.

APPENDIX A

ABBREVIATIONS OF POLYMERS

ABS	Acrylonitrile-butadiene-styrene
AN	Acrylonitrile
CA	Cellulose acetate
CAB	Cellulose acetate butyrate
CAP	Cellulose acetate propionate
CN	Cellulose nitrate
CP	Cellulose propionate
CPE	Chlorinated polyethylene
CPVC	Chlorinated polyvinyl chloride
CTFE	Chlorotrifluoroethylene
DAP	Diallylphthalate
EC	Ethyl cellulose
ECTFE	Poly(ethylene-chlorotrifluoroethylene)
EP	Epoxy
EPDM	Ethylene-propylene-diene monomer
EPR	Ethylene propylene rubber
EPS	Expanded polystyrene
ETFE	Ethylene/tetrafluoroethylene copolymer
EVA	Ethylene-vinyl acetate
FEP	Perfluoro (ethylene-propylene) copolymer
FRP	Fiberglass-reinforced polyester
HDPE	High-density polyethylene
HIPS	High-impact polystyrene
HMWPE	High molecular weight polyethylene
LDPE	Low-density polyethylene
MF	Melamine-formaldehyde
PA	Polyamide
PAPI	Polymethylene polyphenyl isocyanate
PB	Polybutylene
PBT	Polybutylene terephthalate (thermoplastic polyester)
PC	Polycarbonate
PE	Polyethylene
PES	Polyether sulfone
PET	Polyethylene terephthalate
PF	Phenol-formaldehyde
PFA	Polyfluoro-alkoxy
PI	Polyimide
PMMA	Polymethyl methacrylate
PP	Polypropylene

PPO	Polyphenylene oxide
PS	Polystyrene
PSO	Polysulfone
PTFE	Polytetrafluoroethylene
PTMT	Polytetramethylene terephthalate (thermoplastic polyester)
PU	Polyurethane
PVA	Polyvinyl alcohol
PVAC	Polyvinyl acetate
PVC	Polyvinyl chloride
PVDC	Polyvinylidene chloride
PVDF	Polyvinylidene fluoride
PVF	Polyvinyl fluoride
TFE	Polytetrafluoroethylene
SAN	Styrene-acrylonitrile
SI	Silicone
TP	Thermoplastic elastomers
TPX	Polymethylpentene
UF	Urea formaldehyde
UHMWPE	Ultrahigh molecular weight polyethylene
UPVC	Unplasticized polyvinyl chloride

APPENDIX B

GLOSSARY OF POLYMERS AND TESTING

Abrasion resistance: Ability of material to withstand mechanical action such as rubbing, scraping, or erosion that tends to progressively remove material from its surface.

Accelerated aging: Test in which conditions are intensified in order to reduce the time required to obtain a deteriorating effect similar to one resulting from normal service conditions.

Accelerated weathering: Test in which the normal weathering conditions are accelerated by means of a device.

Aging: Process of exposing plastics to natural or artificial environmental conditions for a prolonged period of time.

Amorphous polymers: Polymeric materials that have no definite order or crystallinity. Polymer molecules are arranged in completely random fashion.

Apparent density (bulk density): Weight of unit volume of material including voids (air) inherent in the material.

Arc resistance: Ability of plastic to resist the action of a high voltage electrical arc, usually in terms of time required to render the material electrically conductive.

Birefringence (double refraction): The difference between index of refraction of light in two directions of vibration.

Brittle failure: Failure resulting from inability of material to absorb energy, resulting in instant fracture upon mechanical loading.

Brittleness temperature: Temperature at which plastics and elastomers exhibit brittle failure under impact conditions.

Bulk factor: Ratio of volume of any given quantity of the loose plastic material to the volume of the same quantity of the material after molding or forming. It is a measure of volume change that may be expected in fabrication.

Burst strength: The internal pressure required to break a pressure vessel such as a pipe or fitting. The pressure (and therefore the burst strength) varies with the rate of pressure buildup and the time during which the pressure is held.

Capillary rheometer: Instrument for measuring the flow properties of polymer melts. Comprised of a capillary tube of specified diameter and length, means for applying desired pressures to force molten polymer through the capillary, means for maintaining the desired temperature of the apparatus, and means for measuring differential pressures and flow rates.

Chalking: A whitish, powdery residue on the surface of a material caused by material degradation (usually from weather).

Charpy impact test: A destructive test of impact resistance, consisting of placing the specimen in a horizontal position between two supports, then striking the specimen with a pendulum striker swung from a fixed height. The magnitude of the blow is increased until specimen breaks. The result is expressed in in.-lb or ft-lb of energy.

CIE (Commission Internationale de l'Eclairage): International commission on illuminants responsible for establishing standard illuminants.

Coefficient of thermal expansion: Fractional change in length or volume of a material for unit change in temperature.

Colorimeter: Instrument for matching colors with results approximately the same as those of visual inspection, but more consistently.

Compressive strength: Maximum load sustained by a test specimen in a compressive test divided by original cross-seciton area of the specimen.

Conditioning: Subjecting a material to standard environmental and/or stress history prior to testing.

Continuous use temperature: Maximum temperature at which material may be subjected to continuous use without fear of premature thermal degradation.

Crazing: Undesirable defect in plastic articles, characterized by distinct surface cracks or minute, frost-like internal cracks, resulting from stresses within the article. Such stresses result from molding shrinkage, machining, flexing, impact shocks, temperature changes or action of solvents.

Creep: Due to viscoelastic nature, a plastic subjected to a load for a period of time tends to deform more than it would from the same load released immediately after application. The degree of this deformation is dependent on the load duration. Creep is the permanent deformation resulting from prolonged application of stress below the elastic limit. Creep at room temperature is called cold flow.

Creep modulus (apparent modulus): Ratio of initial applied stress to creep strain.

Creep rupture strength: Stress required to cause fracture in a creep test.

Cross-linking: The setting up of chemical links between the molecular chains. When extensive, as in most thermosetting resins, cross-linking makes one infusible supermolecule of all the chains. Cross-linking can be achieved by irradiation with high energy electron beams or by chemical cross-linking agents.

Crystallinity: State of molecular structure attributed to existence of solid crystals with a definite geometric form. Such structures are characterized by uniformity and compactness.

Cup flow test: Test for measuring the flow properties of thermosetting materials. A standard mold is charged with preweighed material, and the mold is closed using sufficient pressure to form a required cup. Minimum pressures required to mold a standard cup and the time required to close the mold fully are determined.

Cup viscosity test: Test for making flow comparisons under strictly comparable conditions. The cup viscosity test employs a cup-shaped gravity device that permits the timed flow of a known volume of liquid passing through an orifice located at the bottom of the cup.

Density: Weight per unit volume of a material expressed in grams per cubic centimeter, pounds per cubic foot, etc.

Dielectric constant (permititivity): Ratio of the capacitance of a given configuration of electrodes with a material as dielectric to the capacitance of the same electrode configuration with a vacuum (or air for most practical purposes) as the dielectric.

Dielectric strength: Electric voltage gradient at which an insulating material is broken down or ''arced through'' in volts per mil of thickness.

Differential scanning calorimetry (DSC): Thermal analysis technique that measures the quantity of energy absorbed or evolved (given off by a specimen in calories as its temperature is changed).

Dimensional stability: Ability to retain the precise shape in which it was molded, fabricated, or cast.

Dissipation factor: Ratio of the conductance of a capacitor in which the material is dielectric to its susceptance, or the ratio of its parallel reactivity to its parallel resistance. Most plastics have a low dissipation factor, a desirable property because it minimizes the waste of electrical energy as heat.

Drop impact test: Impact resistance test in which a predetermined weight is allowed to fall freely onto the specimen from varying heights. The energy absorbed by the specimen is measured and expressed in in.-lb or ft-lb.

Ductility: Extent to which a material can sustain plastic deformation without fracturing.

Durometer hardness: Measure of the indentation hardness of plastics. It is the extent to which a spring-loaded steel indentor protrudes beyond the pressure foot into the material.

Elongation: The increase in length of a test specimen produced by a tensile load. Higher elongation indicates higher ductility.

Embrittlement: Reduction in ductility due to physical or chemical changes.

Environmental stress cracking: The susceptibility of a thermoplastic article to crack or craze formation under the influence of certain chemicals and stress.

Extensometer: Instrument for measuring changes in linear dimensions (also called strian gauge).

Extrusion plastometer (rheometer): A type of viscometer used for determining the melt index of a polymer. Comprised of a vertical cylinder with two longitudinal bored holes (one for measuring temperature and one for containing the specimen, the latter having an orifice of stipulated diameter at the bottom and a plungering from the top). The cylinder is heated by external bands and weight is placed on the plunger to force the polymer specimen through the orifice. The result is reported in g/10 min.

Fadometer: An apparatus for determining the resistance of materials to fading by exposing them to ultraviolet rays of approximately the same wavelength as those found in sunlight.

Fatigue failure: The failure or rupture of a plastic under repeated cyclic stress, at a point below the normal static breaking strength.

Fatigue limit: The stress below which a material can be stressed cyclically for an infinite number of times without failure.

Fatigue strength: The maximum cyclic stress a material can withstand for a given number of cycles before failure.

Flammability: Measure of the extent to which a material will support combustion.

Flexural modulus: Ratio of the applied stress on a test specimen in flexure to the corresponding strain in the outermost fiber of the specimen. Flexural modulus is the measure of relative stiffness.

Flexural strength: The maximum stress in the outer fiber at the moment of crack or break.

Foamed plastics (cellular plastics): Plastics with numerous cells disposed throughout its mass. Cells are formed by a blowing agent or by the reaction of the constituents.

Gel permeation chromatography (GPC): Column chromatography technique employing a series of columns containing closely packed rigid gel particles. The polymer to be analyzed is introduced at the top of the column and then is eluted with a solvent. The polymer molecules diffuse through the gel at rates depending on their molecular size. As they emerge from the columns, they are detected by differential refractometer coupled to a chart recorder, on which a molecular weight distribution curve is plotted.

Gel point: The stage at which liquid begins to gel, i.e., exhibits pseudoelastic properties.

Hardness: The resistance of plastic materials to compression and indentation. Brinnel hardness and shore hardness are major methods for testing this property.

Haze: The cloudy or turbid aspect of appearance of an otherwise transparent specimen caused by light scattered from within the specimen or from its surface.

Hooke's law: Stress is directly proportional to strain.

Hoop stress: The circumferential stress in a material of cylindrical form subjected to internal or external pressure.

Hygroscopic: Material having the tendency to absorb moisture from air. Plastics, such as nylons and ABS, are hygroscopic and must be dried prior to molding.

Hysteresis: The cyclic noncoincidence of the elastic loading and the unloading curves under cyclic stressing. The area of the resulting elliptical hysteresis loop is equal to the heat generated in the system.

Impact strength: Energy required to fracture a specimen subjected to shock.

Impact test: Method of determining the behavior of material subjected to shock loading in bending or tension. The quantity usually measured is the energy absorbed in fracturing the specimen in a single blow.

Indentation hardness: Resistance of a material to surface penetration by an indentor. The hardness of a material as determined by the size of an indentation made by an indenting tool under a fixed load, or the load necessary to produce penetration of the indentor to a predetermined depth.

Index of refraction: Ratio of velocity of light in vacuum (or air) to its velocity in a transparent medium.

Infrared analysis: Techique used for polymer identification. An infrared spectrometer directs infrared radiation through a film or layer of specimen and measures the relative amount of energy absorbed by the specimen as a function of wavelength or frequency of infrared radiation. The chart produced is compared to correlation charts for known substances to identify the specimen.

Inherent viscosity: In dilute solution viscosity measurements, inherent viscosity is the ratio of the natural logarithm of the relative viscosity to the concentration of the polymer in g/100 ml of solvent.

Intrinsic viscosity: In dilute solution viscosity measurements, intrinsic viscosity is the limit of the reduced and inherent viscosities as the concentration of the polymeric solute approaches zero and represents the capacity of the polymer to increase viscosity.

ISO: Abbreviation for the International Standards Organization.

Isochronous (equal time) stress-strain curve: A stress-strain curve obtained by plotting the stress vs. corresponding strain at a specific time of loading pertinent to a particular application.

Izod impact test: Method for determining the behavior of materials subjected to shock loading. Specimen supported as a cantilever beam is struck by a weight at the end of a pendulum. Impact strength is determined from the amount of energy required to fracture the specimen. The specimen may be notched or unnotched.

Melt index test: Melt index (MI) test measures the rate of extrusion of a thermoplastic material through an orifice of specific length and diameter under prescribed conditions of temperature and pressure. Value is reported in g/10 min for specific condition.

Modulus of elasticity (elastic modulus, Young's modulus): The ratio of stress to corresponding strain below the elastic limit of a material.

Molecular weight: The sum of the atomic weights of all atoms in a molecule. In high polymers, the molecular weight of individual molecules varies widely, therefore, they are expressed as weight average or number average molecular weight.

Molecular weight distribution (MWD): The relative amount of polymers of different molecular weights that comprise a given specimen of a polymer.

Monomer (monomer single-unit): A relatively simple compound that can react to form a polymer (multiunit) by combination with itself or with other similar molecules or compounds.

Necking: The localized reduction in cross-section that may occur in a material under stress. Necking usually occurs in a test bar during a tensile test.

Notch sensitivity: Measure of reduction in load-carrying ability caused by stress concentration in a specimen. Brittle plastics are more notch sensitive than ductile plastics.

Orientation: The alignment of the crystaline structure in polymeric materials so as to produce a highly uniform structure.

Oxygen index: The minimum concentration of oxygen expressed as a volume percent, in a mixture of oxygen and nitrogen that will just support flaming combustion of a material initially at room temperature under the specified conditions.

Peak exothermic temperature: The maximum temperature reached by reacting thermosetting plastic composition is called peak exothermic temperature.

Photoelasticity: Experimental technique for the measurement of stresses and strains in material objects by means of the phenomenon of mechanical birefringence.

Poisson's ratio: Ratio of lateral strain to axial strain in an axial-loaded specimen. It is a constant that relates the modulus of rigidity to Young's modulus.

Polarized light: Polarized electromagnetic radiation whose frequency is in the optical region.

Polarizer: A medium or a device used to polarize the incoherent light.

Polymerization: A chemical reaction in which the molecules of monomers are linked together to form polymers.

Proportional limit: The greatest stress that a material is capable of sustaining without deviation from proportionality of stress and strain (Hooke's law).

Relative viscosity: Ratio of kinematic viscosity of a specified solution of the polymer to the kinematic viscosity of the pure solvent.

Rheology: The science dealing with the study of material flow.

Rockwell hardness: Index of indentation hardness measured by a steel ball indentor.

Secant modulus: The ratio of total stress to corresponding strain at any specific point on the stress-strain curve.

Shear rate: The overall velocity over the cross-section of a channel with which molten or fluid layers are gliding along each other or along the wall in laminar flow.

Shear strength: The maximum load required to shear a specimen in such a volume manner that the resulting pieces are completely clear of each other.

Shear stress: The stress developing in a polymer melt when the layers in a cross-section are gliding along each other or along the wall of the channel (in laminar flow).

SPE: Abbreviation for Society of Plastics Engineers.

Specfic gravity: The ratio of the weight of the given volume of a material to that of an equal volume of water at a stated temperature.

Spectrophotometer: An instrument that measures transmission or apparent reflectance of visible light as a function of wavelength, permitting accurate analysis of color or accurate comparison of luminous intensities of two sources of specific wavelengths.

Specular gloss: The relative luminous reflectance factor of a specimen at the specular direction.

SPI: Abbreviation for Society of Plastics Industry.

Spiral flow test: A method for determining the flow properties of a plastic material based on the distance it will flow under controlled conditions, pressure, and temperature along the path of a spiral cavity using a controlled charge.

Strain: The change in length per unit of original length, usually expressed in percent.

Stress: The ratio of applied load of the original cross-sectional area expressed in pounds per square inch.

Stress concentration: The magnification of the level of applied stress in the region of a notch, crack, void, inclusion, or other stress risers.

Stress optical sensitivity: The ability of materials to exhibit double refraction of light when placed under stress.

Stress relaxation: The gradual decrease in stress with time under a constant deformation (strain).

Stress-strain diagram: Graph of stress as a function of strain. It is constructed from the data obtained in any mechanical test in which a load is applied to a material and continuous measurements of stress and strain are made simultaneously.

Tensile impact energy: The energy required to break a plastic specimen in tension by a single swing of a calibrated pendulum.

Tensile strength: Ultimate strength of a material subjected to tensile loading.

Thermal conductivity: The ability of a material to conduct heat. The coefficient of thermal conductivity is expressed as the quantity of heat that passes through a unit cube of the substance in a given unit of time when the difference in temperature of the two faces is 1°.

Thermogravimetric analysis (TGA): A testing procedure in which changes in the weight of a specimen are recorded as the specimen is progressively heated.

Thermomechanical analysis (TMA): A thermal analysis technique consisting of measuring physical expansion or contraction of a material or changes in its modulus or viscosity as a function of temperature.

Thermoplastic: A class of plastic material that is capable of being repeatedly softened by heating and hardened by cooling. ABS, PVC, polystyrene, polyethylene, etc., are thermoplastic materials.

Thermosetting plastics: A class of plastic materials that will undergo a chemical reaction by the action of heat, pressure, catalysts, etc., leading to a relatively infusible, nonreversible state. Phenolics, epoxies, and alkyds are examples of typical thermosetting plastics.

Torsion: Stress caused by twisting a material.

Torsion pendulum: Equipment used for determining dynamic mechanical plastics.

Toughness: The extent to which a material absorbs energy without fracture. The area under a stress-strain diagram is also a measure of toughness of a material.

Tristimulus colorimeter: The instrument for color measurement based on spectral tristimulus values. Such an instrument measures color in terms of three primary colors: red, green, and blue.

Ultrasonic testing: A nondestructive testing technique for detecting flaws in material and measuring thickness based on the use of ultrasonic frequencies.

Ultraviolet: The region of the electromagnetic spectrum between the violet end of visible light and the X-ray region, including wavelengths from 100 to 3900 Å. Photon of radiations in the UV area have sufficient energy to initiate some chemical reactions and to degrade some plastics.

Vicat softening point: The temperature at which a flat-ended needle of 1 mm^2 circular or square cross-section will penetrate a thermoplastic specimen to a depth of 1 mm under a specified load using a uniform rate of temperature rise.

Viscometer: An instrument used for measuring the viscosity and flow properties of fluids.

Viscosity: A measure of resistance of flow due to internal friction when one layer of fluid is caused to move in relationship to another layer.

Water absorption: The amount of water absorbed by a polymer when immersed in water for stipulated periods of time.

Weathering: A term encompassing exposure of polymers to solar or ultraviolet light, temperature, oxygen, humidity, snow, wind, pollution, etc.

Weatherometer: An instrument used for studying the effect of weather on plastics in an accelerated manner using artificial light sources and simulated weather conditions.

Yellowness index: Measure of the tendency of plastics to turn yellow upon long-term exposure to light.

Yield point: Stress at which strain increases without accompanying increase in stress.

Yield strength: The stress at which a material exhibits a specified limiting deviation from the proportionality of stress to strain. Unless otherwise specified, this stress will be the stress at the yield point.

Young's modulus: The ratio of tensile stress to tensile strain below the proportional limit.

APPENDIX C

DESCRIPTION OF PROFESSIONAL AND TESTING ORGANIZATIONS

AMERICAN NATIONAL STANDARDS INSTITUTE (ANSI)

ANSI is a federation of standards competents from commerce and industry, professional, trade, consumer and labor organizations, and government. ANSI helps to perform the following:

- Identifies the needs for standards and sets priorities for their completion.
- Assigns development work to competent and willing organizations.
- Sees to it that public interests, including those of the consumer, are protected and represented.
- Supplies standards writing organizations with effective procedures and management services to ensure efficient use of their manpower and financial resources and timely development of standards.
- Follows up to assure that needed standards are developed on time.

Another role is to approve standards as American National Standards when they meet consensus requirements. It approves a standard only when it has verified evidence presented by a standards developer that those affected by the standard have reached substantial agreement on its provisions. ANSIs other major roles are to represent U.S. interests in nongovernmental international standards work, to make national and international standards available, and to inform the public.

AMERICAN SOCIETY FOR TESTING AND MATERIALS (ASTM)

ASTM is a scientific and technical organization formed for "the development of standards on characteristics and performance of materials, products, systems and services and the promotion of related knowledge". ASTM is the world's largest source of voluntary consensus standards. The society operates through more than 135 main technical committees with 1550 subcommittees. These committees function in prescribed fields under regulations that ensure balanced representation among producers, users, and general interest participants. The society currently has 28,000 active members, of whom approximately 17,000 serve as technical experts on committees, representing 76,200 units of participation.

Membership in the society is open to all concerned with the fields in which ASTM is active. An ASTM standard represents a common viewpoint of those parties concerned with its provisions, namely, producers, users, and general interest groups. It is intended to aid industry, government agencies, and the general public. The use of an ASTM standard is voluntary. It is recognized that, for certain work, ASTM specifications may be either more or less restrictive than needed. The existence of an ASTM standard does not preclude anyone from manufacturing, marketing, or purchasing products, or using products, processes, or procedures not conforming to the standard. Because ASTM standards are subject to periodic reviews and revision, it is recommended that all serious users obtain the latest revision. A new edition of the *Book of Standards* is issued annually. On the average about 30% of each part is new or revised.

FOOD AND DRUG ADMINISTRATION (FDA)

The Food and Drug Administration is a U.S. government agency of the Department of Health and Human Services. The FDAs activities are directed toward protecting the health of the nation against impure and unsafe foods, drugs, and cosmetics and other potential hazards.

The plastics industry is mainly concerned with the Bureau of Foods, which conducts research and develops standards on the composition, quality, nutrition, and safety of foods, food additives, colors and cosmetics, and conducts research designed to improve the detection, prevention, and control of contamination. The FDA is concerned about indirect additives. Indirect additives are those substances capable of migrating into food from contacting plastic materials. Extensive tests are carried out by the FDA before issuing safety clearance to any plastic material that is to be used in food contact applications. Plastics used in medical devices are tested with extreme caution by the FDAs Bureau of Medical Devices which develops FDA policy regarding safety and effectiveness of medical devices.

NATIONAL BUREAU OF STANDARDS (NBS)

National Bureau of Standards' overall goal is to strengthen and advance the nation's science and technology and to facilitate their effective application for public benefit.

The bureau conducts research and provides a basis for the nation's physical measurement system, scientific and technological services for industry and government, a technical basis for increasing productivity and innovation, promoting international competitiveness in American industry, maintaining equity in trade and technical services, promoting public safety. The bureau's technical work is performed by the National Measurement Laboratory, the National Engineering Laboratory, and the Institute for Computer Sciences and Technology.

NATIONAL ELECTRICAL MANUFACTURERS ASSOCIATION (NEMA)

The National Electrical Manufacturers Association consists of manufacturers of equipment and apparatus for the generation, transmission, distribution, and utilization of electric power. The membership is limited to corporations, firms, and individuals actively engaged in the manufacture of products included within the product scope of NEMA product subdivisions.

NEMA develops product standards covering such matters as nomenclature, ratings, performance, testing, and dimensions. NEMA is also actively involved in developing National Electrical Safety Codes and advocating their acceptance by state and local authorities. Along with a monthly news bulletin, NEMA also publishes manuals, guidebooks, and other material on wiring, installation of equipment, lighting, and standards. The majority of NEMA standardization activity is in cooperation with other national organizations. The manufacturers of wires and cables, insulating materials, conduits, ducts, and fittings are required to adhere to NEMA standards by state and local authorities.

NATIONAL FIRE PROTECTION ASSOCIATION (NFPA)

The National Fire Protection Association has the objective of developing, publishing, and disseminating standards intended to minimize the possibility and effect of fire and explosion. NFPAs membership consists of individuals from business and industry, fire service, health care, insurance, educational, and government institutions. NFPA conducts fire safety education programs for the general public and provides information on fire protection and prevention. Also provided by the association is the field service by specialists on flammable liquids, electricity, gases, and marine problems.

Each year, statistics on causes and occupancies of fires and deaths resulting from fire are compiled and published. NFPA sponsors seminars on the Life Safety Codes, National Electrical Code, industrial fire protection, hazardous materials, transportation emergencies, and other related topics. NFPA also conducts research programs on delivery systems for public fire protection, arson, residential fire sprinkler systems, and other subjects. NFPA publications include *National Fire Codes Annual, Fire Protection Handbook, Fire Journal,* and *Fire Technology.*

NATIONAL SANITATION FOUNDATION (NSF)

The National Sanitation Foundation is an independent, nonprofit environmental organization of scientists, engineers, technicians, educators, and analysts. NSF frequently serves as a trusted neutral agency for government, industry, and consumers, helping them to resolve differences and unite in achieving solutions to problems of the environment.

At NSF, a great deal of work is done on the development and implementation of NSF standards and criteria for health-related equipment. The majority of NSF standards relate to water treatment and purification equipment, products for swimming pool applications, plastic pipe for potable water as well as drain, waste and vent (DWV) uses, plumbing components for mobile homes and recreational vehicles, laboratory furniture, hospital cabinets, polyethylene refuse bags and containers, aerobic waste treatment plants, and other products related to environmental quality.

Manufacturers of equipment, materials, and products that conform to NSF standards are included in official listings and these producers are authorized to place the NSF seal on their products. Representatives from NSF regularly visit the plants of manufacturers to make certain that products bearing the NSF seal fulfill applicable NSF standards.

PLASTICS TECHNICAL EVALUATION CENTER (PLASTEC)

PLASTEC is one of 20 information analysis centers sponsored by the Department of Defense to provide the defense community with a variety of technical information services applicable to plastics, adhesives, and organic matrix composites. For the last 21 years, PLASTEC has served the defense community with authoritative information and advice in such forms as engineering assistance, responses to technical inquiries, special investigations, field troubleshooting, failure analysis, literature searches, state-of-the-art reports, data compilations, and handbooks. PLASTEC has also been heavily involved in standardization activities. In recent years, PLASTEC has been permitted to serve private industry.

The significant difference between a library and technical evaluation center is the quality of the information provided to the user. PLASTEC uses its database library as a means to an end to provide succinct and timely information which has been carefully evaluated and analyzed. Examples of the activity include recommendation of materials, counseling on designs, and performing tradeoff studies between various materials, performance requirements, and costs. Applications are examined consistent with current manufacturing capabilities, and the market availability of new and old materials alike is considered. PLASTEC specialists can reduce raw data to the user's specifications and supplement them with unpublished information that updates and refines published data. PLASTEC works to spin-off the results of government-sponsored R & D to industry and similarly to utilize commercial advancements to the government's goal of highly sought technology transfer. PLASTEC has a highly specialized library to serve the varied needs of their own staff and customers.

PLASTEC offers a great deal of information and assistance to the design engineer in the area of specifications and standards on plastics. PLASTEC has a complete visual search microfilm file and can display and print the latest issues of specifications, test methods, and standards from the U.K., Germany, Japan, U.S., and the International Standards Organization. Military and federal specifications and standards and industry standards such as ASTM, NEMA, and UL are on file and can be quickly retrieved.

SOCIETY OF PLASTICS ENGINEERS (SPE)

The Society of Plastics Engineers promotes scientific and engineering knowledge relating to plastics. SPE is a professional society of plastics scientists, engineers, educators, students, and others interested in the design, development, production, and utilization of plastics materials, products, and equipment. SPE currently has over 22,000 members scattered among its 80 sections. The individual sections as well as the SPE main body arranges and conducts monthly meetings, conferences, educational seminars, and plant tours throughout the year. SPE also publishes *Plastics Engineering, Polymer Engineering and Science, Plastics Composites,* and the *Journal of Vinyl Technology.* The society presents a number of awards each year encompassing all levels of the organization, section, division, committee, and international. SPE divisions of interest are color and appearance, injection molding, extrusion, electrical and electronics, thermoforming, engineering properties and structure, vinyl plastics, blow molding, medical plastics, plastics in building, decorating, mold making, and mold design.

THE SOCIETY OF PLASTICS INDUSTRY (SPI)

The Society of Plastics Industry is a major society, whose membership consists of manufacturers and processors of plastics materials and equipment. The society has four major operating units consisting of the Eastern Seciton, the Midwest Section, the New England Section, and the Western Section. SPIs Public Affairs Committee concentrates on coordinating and managing the response of the plastics industry to issues such as toxicology, combustibility, solid waste, and energy. The Plastic Pipe Institute is one of the most active divisions, promoting the proper use of plastic pipes by establishing standards, test procedures, and specifications. Epoxy Resin Formulators Division has published over 30 test procedures and technical specifications. Risk management, safety standards, productivity, and quality are a few of the major programs undertaken by the machinery division. SPIs other divisions include Expanded Polystyrene Division, Fluoropolymers Division, Furniture Division, Inernational Division, Plastic Bottle Institute, Machinery Division, Molders Division, Mold Makers Division, Plastic Beverage Container Division, Plastic Packaging Strategy Group, Polymeric Materials Producers Division, Polyurethane Division, Reinforced Plastic/Composites Institute, Structural Foam Division, Vinyl Siding Institute, and Vinyl Formulators Division.

The National Plastics Exposition and Conference, held every 3 years by the Society of Plastic Industry, is one of the largest plastic shows in the world.

UNDERWRITERS LABORATORIES (UL)

Underwriters Laboratories, is a not-for-profit organization whose goods are to establish, maintain, and operate laboratories for the investigation of materials, devices, products equipment, constructions, methods, and systems with respect to hazards affecting life and property.

UL has five testing facilities in the U.S. and over 200 inspection centers. More than 700 engineers and 500 inspectors conduct tests and follow-up investigations to ensure that potential hazards are evaluated and proper safeguards provided. UL has six basic services it offers to manu-facturers, inspection authorities, or government officials: product listing service, classification, service, component recognition service, certificate service, inspection service, and fact finding and research.

ULs Electrical Department is in charge of evaluating individual plastics and other products using plastics as components. The Electrical Department evaluates consumer products such as TV sets, power tools, appliances, and industrial and commercial electrical equipment and components. In order for a plastic material to be recognized by UL it must pass a variety of UL tests, including the UL 94 flammability test and the UL 746 series, short and long-term property evaluation tests. When a plastic material is granted Recognized Component Status, a yellow card is issued. The card contains precise iden-tification of the material including supplier, product designation, color, and its UL 94 flammability classification at one or more thicknesses. Also included are many of the property values such as temperature index, hot wire ignition, high-current arc ignition and arc resistance. These data also appear in the recognized component directory.

UL publishes the names of the companies that have demonstrated the ability to provide a product conforming to the established requirements, upon successful completion of the investigation and after agreement of the terms and condition of the listing and follow-up service. Listing signifies that pro-duction samples of the product have been found to comply with the require-ments and that the manufacturer is authorized to use the UL listing mark on the listed products that comply with the requirements.

ULs consumer advisory council was formed to advise UL in establishing levels of safety for consumer products to provide UL with additional user field experience and failure information in the field of product safety, and to aid in educating the general public in the limitations and safe use of specific consumer products.

INDEX